Library of
Davidson College
VOID

EXACT PHILOSOPHY

SYNTHESE LIBRARY

MONOGRAPHS ON EPISTEMOLOGY,

LOGIC, METHODOLOGY, PHILOSOPHY OF SCIENCE,

SOCIOLOGY OF SCIENCE AND OF KNOWLEDGE,

AND ON THE MATHEMATICAL METHODS OF

SOCIAL AND BEHAVIORAL SCIENCES

Editors:

DONALD DAVIDSON, *The Rockefeller University and Princeton University*

JAAKKO HINTIKKA, *Academy of Finland and Stanford University*

GABRIËL NUCHELMANS, *University of Leyden*

WESLEY C. SALMON, *Indiana University*

EXACT PHILOSOPHY

Problems, Tools, and Goals

Edited by

MARIO BUNGE

Foundations and Philosophy of Science Unit, McGill University, Montreal

D. REIDEL PUBLISHING COMPANY

DORDRECHT-HOLLAND / BOSTON-U.S.A.

Library of Congress Catalog Card Number 72-77872

ISBN 90 277 0251 9

Published by D. Reidel Publishing Company,
P.O. Box 17, Dordrecht, Holland

Sold and distributed in the U.S.A., Canada and Mexico
by D. Reidel Publishing Company, Inc.
306 Dartmouth Street, Boston,
Mass. 02116, U.S.A.

All Rights Reserved
Copyright © 1973 by D. Reidel Publishing Company, Dordrecht, Holland
No part of this book may be reproduced in any form, by print, photoprint, microfilm,
or any other means, without written permission from the publisher

Printed in The Netherlands by D. Reidel, Dordrecht

FOREWORD

The papers that follow were read and discussed at the first Symposium on Exact Philosophy. This conference was held at Montreal on November 4th and 5th, 1971, to celebrate the sesquicentennial of McGill University and establish the Society for Exact Philosophy.

The expression 'exact philosophy' is taken to signify *mathematical philosophy*, i.e., philosophy done with the explicit help of mathematical logic and mathematics. So far the expression denotes an attitude rather than a fully blown discipline: it intends to convey the intention to try and proceed in as exact a manner as we can in formulating and discussing philosophical problems and theories. The kind of philosophy we wish to practice and promote is disciplined rather than wild, systematic rather than disconnected, and capable of being argued over rather than oracular. We believe that even metaphysics, notoriously riotous, can be subjected to the control of logic and mathematics. Even the history of philosophy, notoriously unsystematic, can benefit from an exact reconstruction of some classical ideas.

Exactness, though desirable, should not be taken for an end: it is a means for enhancing clarity and systemicity, hence control. Exactness, whether in philosophy or in science, does not guarantee certainty: it eases the discovery of error and its correction. Nor does exactness ensure depth, hence interest: it warrants the possibility of rational scrutiny. The ideal is, of course, to tackle genuine and deep problems in an exact manner. But before we can solve any deep problems in exact philosophy we must accumulate a stock of modest yet exact theories. I submit then that our immediate task is to build well circumscribed theories in exact philosophy: in semantics, epistemology, metaphysics, value theory, ethics, legal logic, etc. The big systems will emerge, if at all, from the fusion of such modest but exact theories.

The style of research in exact philosophy is the one familiar from mathematics and theoretical science: Finding or inventing a problem / Formulating the problem in a reasonably precise way / Trying possible solutions within available theories or in theories built *ad hoc* / Checking the possible solutions against the body of relevant knowledge / Choosing and weighing a solution / Revising the rest of the relevant knowledge in the light of the previous result. The initial problem may be internal to an existing theory or it may consist in building a new theory. While the former situation is common in the advanced area of exact philosophy, namely logic (whether ordinary or not, truth functional or not), in other areas we are usually faced with the much harder task of building new theories. Some of these theories will formalize (perhaps in alternative ways) philosophical insights inherited from inexact philosophy, while others may have no roots in the past although they may settle questions handed down by traditional philosophy. In either case we should not regard our solutions as perfect: it will be enough if they are perfectible.

The name 'exact philosophy' seems to be new. It was born, or at least revived, in typical Viennese fashion, namely at a restaurant table. To be exact, the birth took place at the Vienna Rathauskeller on the evening of September 6th, 1968, during the XIVth International Congress of Philosophy. The name was proposed by the late mathematician and philosopher Richard Montague (1930–1971), whose untimely death prevented him from reading a paper at our Symposium. However, the denotatum of 'exact philosophy' is not new: it has been going on for quite a while, so that we can now exhibit a number of results in exact philosophy. To begin with there are both the standard and the noncanonical systems of mathematical logic, as well as the semantics of logics and of mathematics, which belong to both mathematics and philosophy. We also have a sprinkling of results in the philosophy of science, in metaphysics, in ethics, and in legal logic, that qualify as members of exact philosophy. But we should acknowledge that most of the exact work in semantics, epistemology, metaphysics, value theory, and ethics, remains to be done. We should take some of the problems inherited from traditional philosophy, as well as the totality of philosophical problems posed by mathematics and science, and should approach them in an exact fashion. If we succeed in handling in an exact way some of the deep problems in these areas we shall be entitled to talk of a new revolution in philosophy.

This Symposium marks the coming of age of the Bertrand Russell Colloquium in Exact Philosophy, launched in February 1970 by our Foundations and Philosophy of Science Unit. I take this opportunity to thank the Canada Council for a generous Killam award that has kept this Unit alive, as well as for funding this Symposium. We are also indebted to the International Union of History and Philosophy of Science (Division of Logic, Methodology and Philosophy of Science) for another generous grant in support of our conference.

MARIO BUNGE

Foundations and Philosophy of Science Unit,
McGill University,
Montreal

CONTENTS

FOREWORD   v

PART I: LOGIC

HUGUES LEBLANC / Matters of Relevance   3
BRIAN F. CHELLAS / Notions of Relevance. Comments on Leblanc's Paper   21

PART II: SEMANTICS

LARS SVENONIUS / Translation and Reduction   31
MARIO BUNGE / A Program for the Semantics of Science   51

PART III: EROTETICS

NUEL D. BELNAP, Jr. / S-P Interrogatives   65

PART IV: PHILOSOPHY OF MATHEMATICS

WILLIAM S. HATCHER / Foundations as a Branch of Mathematics   83
CHARLES CASTONGUAY / Naturalism in Mathematics. Comments on Hatcher's Paper   93

PART V: PHILOSOPHY OF SCIENCE

RAIMO TUOMELA / Deductive Explanation of Scientific Laws   103

PART VI: METAPHYSICS

PETER KIRSCHENMANN / Concepts of Randomness   129

PART VII: ETHICS

BAS C. VAN FRAASSEN / The Logic of Conditional Obligation   151

HARRY BEATTY / On Evaluating Deontic Logics. Comments on
van Fraassen's Paper   173

PART VIII: LEGAL PHILOSOPHY

CARLOS E. ALCHOURRÓN / The Intuitive Background of Normative Legal Discourse and Its Formalization   181

PART IX: HISTORY OF PHILOSOPHY

HÉCTOR-NERI CASTAÑEDA / Plato's *Phaedo* Theory of Relations   201

PART I

LOGIC

HUGUES LEBLANC[1]

MATTERS OF RELEVANCE

Forswearing venerable doctrine, some now claim that the truth-value of a statement A, and – more generally – that of a theory T, hinges upon (the truth-values of) all other statements from the same language as A or T. I should like to investigate this matter as regards languages of four main sorts: first-order languages without a box (called here L^1), first-order languages with a box (called $L^\square$), predicative second-order languages (called L^{2^1}), and impredicative second-order languages (called L^2).

Paraphrasing at times familiar results, I shall report that: (a) when the statements in a theory T come from a language L^1, only some of the atomic wffs of L^1 – those to be known below as *the atomic subformulas of T* – bear on the truth-value of T, and (b) when the statements in T come from a language L^{2^1}, only the atomic wffs of L^{2^1} bear on that truth-value. However, when the statements in T come from a language L^2, some non-atomic wffs of L^2 may affect the truth-value of T, as an example due to Robert K. Meyer and myself attests. The result in (a) can be sharpened somewhat. Indeed, when the statements in T come from a language L^1, the atomic wffs of L^1 that do not figure among the atomic subformulas of T may go truth-valueless, and the truth-value of T will still compute right.

Languages of the sort $L^\square$ will fall into five subfamilies. Those in the first subfamily (called here $L^\square_1$) will have the axioms of I plus the Barcan Formula; those in the second (called $L^\square_2$) will have the axioms of M plus the Barcan Formula; those in the third (called $L^\square_3$) will have the axioms of B; those in the fourth (called $L^\square_4$) will have the axioms of S4 plus the Barcan Formula; and those in the fifth (called $L^\square_5$) will have the axioms of S5.

Breaking possibly new ground, I shall report that, whatever language $L^\square_i$ the statements in a theory T may come from, only the atomic subformulas of T affect the truth-value of T. I shall also note that, when $i=5$, the other atomic wffs of the language may again go truth-valueless and the truth-value of T will nonetheless compute right. Not so, how-

M. Bunge (ed.), Exact Philosophy, 3–20. All Rights Reserved
Copyright © 1973 by D. Reidel Publishing Company, Dordrecht-Holland

ever, when $i<5$. As examples due to Brian Chellas, Robert McArthur, and George Weaver show, a shortage of truth-values may occasionally throw the computation off. To prevent such mishaps, truth-values must be assigned to *all* the atomic wffs of the language *or else* indices must be attached to the truth-value assignments, a trick which allows any assignment to figure more than once in the computation.[2]

Modest though they may be, my results should make for a better documented, and hence more responsible, appraisal of Logical Atomism. The doctrine has its limitations, but – as I show – much of it is fact, not just dogma. They should also throw some fresh light on first-order languages with a box. Some of these, the languages $L^{\square}_5$, behave in a very normal fashion. The rest raise an issue which, in my opinion, Kripke's use of possible worlds simply blurs.

I

I attend in this section to a number of syntactical preliminaries.[3] The *primitive vocabulary* of a language L^1 will consist of:
  (a) $aleph_0$ individual variables,
  (b) $aleph_0$ individual parameters,
  (c) anywhere from 0 to $aleph_0$ individual constants,
  (d) anywhere from 1 to $aleph_0$ predicate constants, each one of them identified as being of degree 1, or degree 2, etc.,
  (e) the three logical operators '$\sim$', '$\supset$', and '$\forall$', and
  (f) the three punctuation signs '(', ')', and ','.
The primitive vocabulary of a language $L^{\square}$ will consist of the signs in (a)–(f) *plus* the logical operator '$\square$'. And that of a language L^2 or $L^{2!}$ will consist of the signs in (a)–(e) *plus* – for each d from 1 on – $aleph_0$ predicate variables of degree d and $aleph_0$ predicate parameters of that degree. As usual, the *formulas* of a language L will be all finite (but nonempty) strings of primitive signs of L.

My variables are in effect what the literature understands by *bound* variables, and my parameters what it understands by *free* variables. Incidentally, the predicate parameters of a language L^2 or $L^{2!}$ will be presumed to come for each degree d in some definite order, to be known as the *alphabetic order* of these parameters.

Because I use both variables and parameters, my arsenal of syntactic variables will be rather large: 'X' to refer to individual variables, '$\mathbf{X}$' to individual parameters and constants, 'I' to *individual signs* in general (i.e., individual variables, individual parameters, and individual constants), 'F' ('F^d' when the degree matters) to predicate variables, '$\mathbf{F}$' ('$\mathbf{F}^d$' when the degree matters) to predicate parameters and constants, 'V' to variables in general, and 'A', 'B', and 'C' to formulas.

Three substitution conventions will be needed as we proceed. Under $C1$, the first of these, $A(\mathbf{X}/X)$ will be the result of replacing everywhere in the formula A the individual variable X by the individual parameter or constant $\mathbf{X}$; and – with F and $\mathbf{F}$ understood to be of the same degree – $A(\mathbf{F}/F)$ will be the result of replacing everywhere in A the predicate variable F by the predicate parameter or constant $\mathbf{F}$.

In all my languages formulas of the following four sorts will count as *wffs* (= well-formed formulas):

(a) $\mathbf{F}^d(\mathbf{X}_1, \mathbf{X}_2, ..., \mathbf{X}_d)$, where $\mathbf{F}^d$ is a predicate constant and $\mathbf{X}_1$, $\mathbf{X}_2, ...,$ and $\mathbf{X}_d$ are individual parameters or constants,

(b) $\sim A$, where A is a wff,

(c) $(A \supset B)$, regularly abridged $A \supset B$, where A and B are wffs, and

(d) $(\forall X) A$, where – for some individual parameter $\mathbf{X} - A(\mathbf{X}/X)$ is a wff.

In languages $L^\square$ formulas of one extra sort besides (a)–(d) will also count as wffs, to wit:

(e) $\square A$, where A is a wff.

And in languages L^2 and L^{21} so will formulas of two extra sorts besides (a)–(d), to wit:

(f) $\mathbf{F}^d(\mathbf{X}_1, \mathbf{X}_2, ..., \mathbf{X}_d)$, where this time around $\mathbf{F}^d$ is a predicate parameter, and

(g) $(\forall F) A$, where – for some predicate parameter $\mathbf{F}$ of the same degree as $F - A(\mathbf{F}/F)$ is a wff.[4]

A few familiar labels will prove handy in what follows. Wffs of the sort $\mathbf{F}^d(\mathbf{X}_1, \mathbf{X}_2, ..., \mathbf{X}_d)$ ($\mathbf{F}^d$ a predicate constant or – when the occasion permits – a predicate parameter, and $\mathbf{X}_1, \mathbf{X}_2, ...,$ and $\mathbf{X}_d$ as in (a) above) will be called *atomic*. Wffs of a language L^2 or L^{21} that do not contain any predicate variable or predicate parameter (and hence qualify as wffs of a language L^1) will be known as *first-order wffs*. Wffs that do not contain any parameter whatever will be known as *statements*, whereas wffs that do will be known as *quasi-statements*. And non-empty sets of statements will be known as *theories*.[5] I shall refer to sets of wffs by means of 'S', and to theories by means of 'T'; and I shall say that a theory T is *(couched) in a language L* if every statement in T comes from L.

Here as in all of the literature, any wff will count as one of its *components*; A will count as a component of $\sim A$, $(\forall X) A$, $\Box A$, and $(\forall F) A$; A and B will count as components of $A \supset B$; and any component of a component of a wff will count as a component of that wff. Here as in Gentzen, any wff will count as one of its *subformulas*; A will count as a subformula of $\sim A$ and $\Box A$; for any individual parameter or constant $\mathbf{X}$, $A(\mathbf{X}/X)$ will count as a subformula of $(\forall X) A$; for any predicate parameter or constant $\mathbf{F}$ of the same degree as F, $A(\mathbf{F}/F)$ will count as a subformula of $(\forall F) A$; and any subformula of a subformula of a wff will count as a subformula of that wff. Finally, any subformula of a member of a theory will count as a subformula of that theory. The atomic subformulas of a theory will star in many of the theorems below.

Now for my last two substitution conventions.

C2. Let A be a wff or a component of one; let $\mathbf{X}_1, \mathbf{X}_2, \ldots$, and $\mathbf{X}_d$ ($d > 0$) be distinct individual parameters; and let $I_1, I_2, \ldots$, and I_d be (not necessarily distinct) individual signs. By $A(I_1, I_2, \ldots, I_d/\mathbf{X}_1, \mathbf{X}_2, \ldots, \mathbf{X}_d)$ I shall understand the result of simultaneously replacing $\mathbf{X}_1$ in A by I_1, $\mathbf{X}_2$ by $I_2, \ldots$, and $\mathbf{X}_d$ by I_d.

C3. Let A be a wff or a component of one; let F^d be a predicate variable of degree d; let $\mathbf{X}_1, \mathbf{X}_2, \ldots$, and $\mathbf{X}_d$ be distinct individual parameters; and let $F^d(I_{1_1}, I_{2_1}, \ldots, I_{d_1})$, $F^d(I_{1_2}, I_{2_2}, \ldots, I_{d_2}), \ldots$, and $F^d(I_{1_k}, I_{2_k}, \ldots, I_{d_k})$ ($k \geqslant 0$) be all the atomic components of A that begin with F^d. *Case 1*: $k = 0$. Then $A(B/F^d(\mathbf{X}_1, \mathbf{X}_2, \ldots, \mathbf{X}_d))$ is to be A. *Case 2*: $k > 0$. If F^d occurs in a component of A of the sort $(\forall V) A'$ and B has a component of the sort $(\forall V) B'$ (V the same variable in both cases), $A(B/F^d(\mathbf{X}_1, \mathbf{X}_2, \ldots, \mathbf{X}_d))$ is to be $A(\mathbf{F}^d/F^d)$, where $\mathbf{F}^d$ is the alphabetically earliest predicate parameter of degree d of the language that A belongs to. Otherwise, $A(B/F^d(\mathbf{X}_1, \mathbf{X}_2, \ldots, \mathbf{X}_d))$ is to be the result of replacing $F^d(I_{1_i}, I_{2_i}, \ldots, I_{d_i})$ everywhere in A by $B(I_{1_i}, I_{2_i}, \ldots, I_{d_i}/\mathbf{X}_1, \mathbf{X}_2, \ldots, \mathbf{X}_d)$, this for each i from 1 through k.

My last assignment in this section is a lengthy one: to specify the circumstances under which a set of wffs of a language L (hence, in particular, a theory in L) is syntactically consistent in L. I discharge it in three steps.

(1) Let A1–A19 be the following *axiom schemata*:

A1. $A \supset (B \supset A)$,
A2. $(A \supset (B \supset C)) \supset ((A \supset B) \supset (A \supset C))$,
A3. $(\sim A \supset \sim B) \supset (B \supset A)$,
A4. $(\forall X)(A \supset B) \supset ((\forall X) A \supset (\forall X) B)$,
A5. $A \supset (\forall X) A$,
A6. $(\forall X) A \supset A(X/X)$, for any individual parameter or constant X,
A7. $(\forall X) A$, where – for some individual parameter X foreign to $(\forall X) A - A(X/X)$ is an axiom,
A8. $(\forall F)(A \supset B) \supset ((\forall F) A \supset (\forall F) B)$,
A9. $A \supset (\forall F) A$,
A10. $(\forall F^d) A \supset A(B/F^d(X_1, X_2, ..., X_d))$, for any wff B and any distinct individual parameters $X_1, X_2 ..., $ and X_d,
A11. Same as A10, but with B restricted to be a first-order wff,
A12. $(\forall F) A$, where – for some predicate parameter F of the same degree as F, but foreign to $(\forall F) A - A(F/F)$ is an axiom,
A13. $\Box(A \supset B) \supset (\Box A \supset \Box B)$,
A14. $\Box A$, where A is an axiom,
A15. $(\forall X) \Box A \supset \Box (\forall X) A$,
A16. $\Box A \supset A$,
A17. $A \supset \Box \Diamond A$ ('$\Diamond$' short here and in A19 for '$\sim \Box \sim$'),
A18. $\Box A \supset \Box \Box A$, and
A19. $\Diamond A \supset \Box \Diamond A$.

The *axioms* of my various languages will then be as indicated in Table I.

TABLE I

L^1: A1–A7
L^2: A1–A10 and A12
$L^{2!}$: A1–A9 and A11–A12
$L_1^\Box$: A1–A7 and A13–A15
$L_2^\Box$: A1–A7 and A13–A16
$L_3^\Box$: A1–A7, A13–A14, and A16–A17
$L_4^\Box$: A1–A7, A13–A16, and A18
$L_5^\Box$: A1–A7, A13–A14, A16, and A19

(A11, which distinguishes $L^{2!}$ form L^2, is sometimes known as the *Predicative Specification Law*; A15 is the *Barcan Formula* mentioned earlier; and A16–A19 are the so-called *characteristic formulas* of M, B, S4, and S5, respectively.)

(2) S being a set of wffs of a language L, and A a wff of L, I shall say that A *is provable in L from S* if there is a finite column of wffs of L which closes with A and in which every entry belongs to S, is an axiom of L, or follows from two previous entries by *Modus Ponens*.

(3) And I shall say that a set S of wffs of a language L is *syntactically consistent in L* if there is no wff A of L such that both A and $\sim A$ are provable in L from S.

II

When it comes to interpreting my various languages, I use truth-value functions rather than models. The account is appropriate, as Theorems 1, 5, and 7 below attest; and it suits my overall purpose to a T.[6] I limit myself in this section to languages without modalities, saving for the next languages with a box.

Understand by *a truth-value function for a language L^1* any function α from the wffs of L^1 to $\{1, 0\}$ such that, for any wff A of L^1,

(i) in case A is a negation $\sim B$, $\alpha(A)=1$ if and only if $\alpha(B)=0$,

(ii) in case A is a conditional $B \supset C$, $\alpha(A)=1$ if and only if $\alpha(B)=0$ or $\alpha(C)=1$, and

(iii) in case A is a quantification $(\forall X) B$, $\alpha(A)=1$ if and only if $\alpha(B(X/X))=1$ for every individual parameter and constant X of L^1.

Next, understand by *a truth-value function for a language L^2* any function α from the wffs L^2 to $\{1, 0\}$ such that, for any wff A of L^2,

(i)–(iii) as before, and

(iv) in case A is a quantification $(\forall F^d) L$, $\alpha(A)=1$ if and only if $\alpha(B(C/F^d(X_1, X_2, \ldots, X_d)))=1$ for every wff C of L^2 and any distinct individual parameters $X_1, X_2, \ldots,$ and X_d of L^2.

Then, understand by a *truth-value function for a language $L^{2!}$* any function α from the wffs of $L^{2!}$ to $\{1, 0\}$ that satisfies conditions (i)–(iv), but with C in (iv) restricted to be a first-order wff. Finally, where α is a truth-value function for a language L^i ($i=1, 2,$ or 2!), and T is a theory in L^i, take $\alpha(T)$ to equal 1 if and only if $\alpha(A)$ equals 1 for every sentence A in T.

Adapting the argument in Henkin (1949), I have shown elsewhere[7] that:

Theorem 1. A theory T in a language L^i ($i=1, 2,$ or $2!$) is syntactically consistent in L^i if and only if $\alpha(T)=1$ for at least one truth-value function α for L^i.[8]

It follows from the theorem that:

(a) when $i=1$, $\alpha(T)=1$ for some truth-value function α for L^i if and only if T has a model,

(b) when $i=2$, $\alpha(T)=1$ for some truth-value function α for L^i if and only if T has what Henkin calls a *general model* satisfying the (unrestricted) Axiom of Specification,[9] and

(c) when $i=2!$, $\alpha(T)=1$ for some truth-value function α for L^i if and only if T has a general model satisfying the Predicative Axiom of Specification.

My semantic account of the languages L^1, L^2, and $L^{2!}$ thus matches the traditional one.

Turning now to questions of relevance, I shall say that, where α is a truth-value function for a language L^i ($i=1, 2,$ or $2!$) and α^* is a function from just the atomic wffs of L^i to $\{1, 0\}$, α has α^* as its *atomic restriction* if α and α^* agree on all the atomic wffs of L^i. And I shall say that *the truth-value of a theory T in L^i depends upon the truth-values of just the atomic wffs of L^i* if, for any two truth-value functions α and α' for L^i that have the same atomic restriction, $\alpha(T)=\alpha'(T)$.

A routine induction will show that, where α is a truth-value function for a language L^1 and α^* is the atomic restriction of α, $\alpha(A)=\alpha^*(A)$ for any wff A of L^1. The induction is on the length of A. Hence:

Theorem 2. The truth-value of a theory T in a language L^1 depends upon the truth-values of just the atomic wffs of L^1.

A second induction will similarly show that, where α is a truth-value function for a language $L^{2!}$ and α^* is the atomic restriction of α, $\alpha(A)=\alpha^*(A)$ for any wff A of $L^{2!}$. The induction is on the number of quantifications in A of the sort $(\forall F)\, B$, with subordinate inductions on the length of A seeing both the basis and the inductive step through. Hence:

Theorem 3. The truth-value of a theory T in a language $L^{2!}$ depends upon the truth-values of just the atomic wffs of $L^{2!}$.

The two results are not unexpected ones, and I include them only for the record.

The story changes, though, when the statements in T come from a language L^2. Meyer and myself have constructed a pair of truth-value functions for a language L^2 that agree on all the atomic wffs of L^2 and yet disagree on the statement '$(\exists f)(\exists x)(\exists y)(f(x) \& \sim f(y))$' ('$f$' here some predicate variable of L^2 of degree 1, and 'x' and 'y' two individual variables of L^2). The result is announced in Leblanc and Meyer (1970), and detailed proof of it can be found in Leblanc and Weaver (1972). So, *the truth-value of a theory T in a language L^2* (unlike that of a theory T in a language L^1 or a language $L^{2!}$) *does not always depend upon the truth-values of just the atomic wffs of L^2*. As in the case of the one-statement theory $\{(\exists f)(\exists x) \supset \supset(\exists y)(f(x) \& \sim f(y))\}$, it may also depend upon the truth-values of various non-atomic wffs of L^2.

As promised on p. 269, Theorem 2 can be sharpened some. T being a theory in a language L^i ($i=1$ or $2!$), α a truth-value function for L^i, and α^* a function from just the atomic subformulas of T to $\{1, 0\}$, I shall say that α has α^* as its *atomic T-restriction* if α and α^* agree on all the atomic subformulas of T. And I shall say that *the truth-value of a theory T in L^i depends upon the truth-values of just the atomic subformulas of T* if, for any two truth-value functions α and α' for L^i that have the same atomic T-restriction, $\alpha(T) = \alpha'(T)$.

A routine induction will show that, where T is a theory in a language L^1, α is a truth-value function for L^1, and α^* is the atomic T-restriction of α, $\alpha(A) = \alpha^*(A)$ for any statement A in T. Hence:

Theorem 4. The truth-value of a theory T in a language L^1 depends upon the truth-values of just the atomic subformulas of T.

Theorem 3, however, cannot be so sharpened. Given any language $L^{2!}$, let α be a truth-value function for $L^{2!}$ that assigns the truth-value 1 to all the atomic wffs of $L^{2!}$, and let α' be one that agrees with α on all the atomic wffs of $L^{2!}$ but one, say: the wff '**f**(x, x)' ('**f**' here some predicate parameter of $L^{2!}$ of degree 2, and 'x' some individual parameter of $L^{2!}$). It is easily verified that the statement '$(\forall f)(\forall x)(\forall y)(f(x) \supset f(y))$' of $L^{2!}$ ('f' here some predicate variable of $L^{2!}$ of degree 1, and 'x' and 'y' two individual variables of $L^{2!}$) has the truth-value 1 under α, but not under α'. So, *the truth-value of a theory T in a language $L^{2!}$* (unlike that of a theory T in a language L^1) *does not always depend upon the truth-values of just the atomic subformulas of T*. As in the case of the one-statement theory $\{(\forall f)(\forall x)(\forall y)(f(x) \supset f(y))\}$, it may also depend upon the truth-values of other atomic wffs of $L^{2!}$.

As Theorems 2–4 suggest, the semantic account on p. 274 of my languages L^1 and $L^{2!}$ can be improved some.

(1) α being a function from the atomic wffs of a language L^i ($i=1$ or 2!) to $\{1, 0\}$, take a wff A of L^i to be *true under* α if:

(i) in case A is atomic, $\alpha(A)=1$,

(ii) in case A is a negation $\sim B$, B is not true under α,

(iii) in case A is a conditional $B \supset C$, B is not true under α or C is,

(iv) in case A is a quantification $(\forall X) B$, $B(X/X)$ is true under α for every individual parameter and constant X of L^i, and

(v) in case A is a quantification $(\forall F^d) B$, $B(C/F^d(X_1, X_2, ..., X_d))$ is true under α for every first-order wff C of L^i and any distinct individual parameters $X_1, X_2, ..., $ and X_d of L^i.

Extant proofs of Theorem 1 can be adapted to show that a theory T in L^i is syntactically consistent in L^i if and only if there is a function from the atomic wffs of L^i to $\{1, 0\}$ under which every member of T is true.

(2) A being a wff of a language L^1, and α being a function from the atomic subformulas of A and possibly other atomic wffs of L^1 to $\{1, 0\}$, take A to be *true under* α if the first four of the conditions in (1) are met. Extant proofs of Theorem 1 can be adapted to show that a theory T in L^1 is syntactically consistent in L^1 if and only if there is a function from the atomic subformulas of T to $\{1, 0\}$ under which every member of T is true.

III

Next for languages with a box. I begin with those labelled $L_5^\square$, as their story is simpler.

Extending the approach in Kripke (1959), understand by a *truth-value pair for a language* $L_5^\square$ any pair of the sort $\langle K, \alpha \rangle$, where K is a non-empty set of functions from the wffs of $L_5^\square$ to $\{1, 0\}$ and α is a member of K. Next, take a wff A of $L_5^\square$ to be *true* under a truth-value pair $\langle K, \alpha \rangle$ for $L_5^\square$ if:

(i) in case A is atomic, $\alpha(A)=1$,

(ii) in case A is a negation $\sim B$, B is not true under $\langle K, \alpha \rangle$,

(iii) in case A is a conditional $B \supset C$, B is not true under $\langle K, \alpha \rangle$ or C is,

(ir) in case A is a quantification $(\forall X) B$, $B(X/X)$ is true under $\langle K, \alpha \rangle$ of every individual parameter and constant X of $L_5^\square$, and

(vv) in case A is a modality $\square B$, B is true under $\langle K, \beta \rangle$ for every β in K.

And, then, take a theory T of $L_5^\square$ to be *true under a truth-value pair* $\langle K, \alpha \rangle$ *for* $L_5^\square$ if every sentence in T is true under $\langle K, \alpha \rangle$.

Adaptation of the argument in Makinson (1969) will show that:

Theorem 5. A theory T in a language $L_5^\square$ is syntactically consistent in $L_5^\square$ if and only if T is true under at least one truth-value pair for $L_5^\square$.

It follows from the theorem that T is true under a truth-value pair for $L_5^\square$ if and only if T has a Kripke model of the kind appropriate for $L_5^\square$,[10] and hence that my semantic account of the languages $L_5^\square$ matches Kripke's.

Generalizing the second of our criteria of truth-relevance, let T be a theory in a language $L_5^\square$, let $\langle K, \alpha \rangle$ be a truth-value pair for $L_5^\square$, let K^* be a non-empty set of functions from the atomic subformulas of the members of T to $\{1, 0\}$, and let α^* be a member of K^*. I shall say that $\langle K, \alpha \rangle$ has $\langle K^*, \alpha^* \rangle$ as *one* of its *atomic T-restrictions* if there is a function f from K^* to K such that:

(a) for any β^* in K^*. β^* and $f(\beta^*)$ agree on all the atomic subformulas of T, and

(b) $f(\alpha^*)$ is α.[11]

And I shall say that *the truth-value of a theory T of $L_5^\square$ depends upon the truth-values of just the atomic subformulas of T* (and, hence, upon those of just the atomic wffs of $L_5^\square$), if, for any two truth-value pairs $\langle K, \alpha \rangle$ and $\langle K', \alpha' \rangle$ for $L_5^\square$ that have the same atomic T-restrictions, T is true under $\langle K, \alpha \rangle$ if and only if T is true under $\langle K', \alpha' \rangle$.

A simple enough induction will show that, where T is a theory in a language $L_5^\square$, $\langle K, \alpha \rangle$ is a truth-value pair for $L_5^\square$, and $\langle K^*, \alpha^* \rangle$ is any atomic T-restriction of $\langle K, \alpha \rangle$, a statement in T is true under $\langle K, \alpha \rangle$ if and only if true under $\langle K^*, \alpha^* \rangle$. Hence:

Theorem 6. The truth-value of a theory T in a language $L_5^\square$ depends upon the truth-values of just the atomic subformulas of T (and, hence, upon those of just the atomic wffs of $L_5^\square$).

First-order theories with an S5 box thus fare exactly like their boxless brethren.

As readers of Kripke will have guessed, treatment of my other modal languages calls for a relation R on K. I supply two accounts. Details of the first, which is simpler but suits only certain theories, are as follows.

Understand by a *truth-value triple for a language* $L_i^\square$ ($i=1, 2, 3,$ or 4) any triple of the sort $\langle K, \alpha, R \rangle$, where K is a non-empty set of functions from the wffs of $L_i^\Box$ to $\{1, 0\}$, α is a member of K, and R is a dyadic relation on K when $i=1$, a reflexive one when $i=2$, a reflexive and symmetrical one when $i=3$, and a reflexive and transitive one when $i=4$. Next, take a wff A of $L_i^\Box$ to be *true under a truth-value triple* $\langle K, \alpha, R \rangle$ *for* $L_i^\Box$ if:

(i) in case A is atomic, $\alpha(A)=1$,

(ii) in case A is a negation $\sim B$, B is not true under $\langle K, \alpha, R \rangle$,

(iii) in case A is a conditional $B \supset C$, B is not true under $\langle K, \alpha, R \rangle$ or C is,

(iv) in case A is a quantification $(\forall X) B$, $B(X/X)$ is true under $\langle K, \alpha, R \rangle$ for every individual parameter and constant X of $L_i^\Box$, and

(v) in case A is a modality $\Box B$, B is true under $\langle K, \beta, R \rangle$ for every B in K such that $R(\alpha, \beta)$.

Then, take a theory T in $L_i^\Box$ to be true under a truth-value triple $\langle K, \alpha, R \rangle$ for $L_i^\Box$ if every sentence in T is true under $\langle K, \alpha, R \rangle$. Finally, call a theory T in $L_i^\Box$ *infinitely extendible* (as regards the atomic wffs of $L_i^\Box$) if aleph$_0$ atomic wffs of $L_i^\Box$ do not figure among the atomic subformulas of T.

Adaptation of the argument in Makinson (1966) will show that:

Theorem 7. An infinitely extendible theory T in a language $L_i^\Box$ ($i=1, 2, 3,$ or 4) is syntactically consistent in $L_i^\Box$ if and only if T is true under at least one truth-value triple for $L_i^\Box$.

The result is a generalization of one I announced at the Temple University Conference on Alternative Semantics, December 29–30, 1970.[12]

Now, generalizing further my second criterion of truth-relevance, let T be a theory in a language $L_i^\Box$ ($i=1, 2, 3,$ or 4), let $\langle K, \alpha, R \rangle$ be a truth-value triple for $L_i^\Box$, let K^*, be a non-empty set of functions from the atomic subformulas of T to $\{1, 0\}$, let α^* be a member of K^*, and let R^* be a relation on K^*. I shall say that $\langle K, \alpha, R \rangle$ has $\langle K^*, \alpha^*, R^*)$ as *one* of its *atomic T-restrictions* if there exists a one-to-one function f from K^* to K such that:

(a) for any β^* in K^*, β^* and $f(\beta^*)$ agree on all the atomic subformulas of T,

(b) $f(\alpha^*)$ is α, and

(c) for any two β^* and γ^* in K^*, $R^*(\beta^*, \gamma^*)$ if and only if $R(f(\beta^*), f(\gamma^*))$.

And I shall say that *the truth-value of a theory T of $L_i^\Box$ depends upon the truth-value of just the atomic subformulas of T* (and, hence, upon those of just the atomic wffs of $L_i^\square$) if, for any two truth-value triples $\langle K, \alpha, R \rangle$ and $\langle K', \alpha', R' \rangle$ for $L_i^\square$ that have the same atomic T-restrictions, T is true under $\langle K, \alpha, R \rangle$ if and only if T is true under $\langle K', \alpha', R' \rangle$.

It is readily checked that the relation R^* in any atomic T-restriction $\langle K^*, \alpha^*, R^* \rangle$ of a truth-value triple $\langle K, \alpha, R \rangle$ for a language $L_i^\square$ is reflexive when R is, symmetrical when R is, etc. So, as one more induction will show, a statement in T is true under $\langle K, \alpha, R \rangle$ if and only if true under $\langle K^*, \alpha^*, R^* \rangle$. Hence:

Theorem 8. The truth-value of an infinitely extendible theory T in a language $L_i^\square$ ($i = 1, 2, 3$, or 4) depends upon the truth-values of just the atomic subformulas of T (and, hence, upon those of just the atomic wffs of $L_i^\square$).

Do *all* modal first-order theories, then, behave alike? Puzzlingly enough, the answer is *No*, as relativizing the notion of a truth-value pair and that of a truth-value triple will show.

T being a theory in a language $L_5^\square$, understand by a *truth-value T-pair for* $L_5^\square$ any pair of the sort $\langle K, \alpha \rangle$, where K is a non-empty set of functions $f$rom the atomic subformulas of T to $\{1, 0\}$, and α is a member of K.

Then take a member A of T to be true under such a pair if conditions (i)–(v) on p. 277 are met, and take T to be true under the pair if every member of T is. In view of Theorem 6 one might expect that T is syntactically consistent in $L_5^\square$ if and only if T is true under some truth-value T-pair or other for $L_5^\square$, and inspection of the proof of Theorem 6 bears this out:

Theorem 9. A theory T in a language $L_5^\square$ is syntactically consistent in $L_5^\square$ if and only if T is true under at least one truth-value T-pair for $L_5^\square$.

As a result, the semantic account on pp. 277–8 of my languages $L_5^\square$ can be improved some. A being a wff of $L_5^\square$, K being a non-empty set of functions from the atomic subformulas of A and possibly other atomic wffs of $L_5^\square$ to $\{1, 0\}$, and α being a member of K, take A to be true under $\langle K, \alpha \rangle$ if conditions (i)–(v) on p. 277 are met. And, T being a theory in $L_5^\square$, K being a non-empty set of functions from the atomic subformulas of T to $\{1, 0\}$ and α being a member of K, take T to be true under $\langle K, \alpha \rangle$ if every member of T is true under $\langle K, \alpha \rangle$. This thriftier account, incidentally, makes for a decision procedure when T has only finitely many atomic subformulas.

Now for truth-value triples. T being this time a theory in a language $L_i^\square$ ($i = 1, 2, 3$, or 4), understand by a *truth-value T-triple for* $L_i^\square$ any triple of the sort $\langle K, \alpha, R \rangle$, where K is a non-empty set of functions from the atomic subformulas of T to $\{1, 0\}$, α is a member of K, and R is a dyadic relation on K when $i=1$, a reflexive one when $i=2$, etc. Then take a member A of T to be true under such a triple if conditions (i)–(v) on p. 279 are met, and take T to be true under the triple if every member of T is.

Theorem 8 notwithstanding, an infinitely extendible theory T in $L_i^\square$ ($i<5$) may be syntactically consistent in $L_i^\square$ and yet fail to be true under any truth-value T-triple for $L_i^\square$. Cases in point are as follows.

(a) The theory $\{\Diamond\square f(a), \sim\square\square f(a), \sim\Diamond\Diamond f(a)\}$, though syntactically consistent in $L_1^\square$, is not true under any truth-value T-triple for $L_1^\square$. I owe the example to Chellas.[13]

(b) The theory $\{\square f(a), \sim\square\square f(a)\}$, though syntactically consistent in $L_2^\square$, is not true under any truth-value T-triple for $L_2^\square$, and this for a simple enough reason: any reflexive relation on the sets $\{\alpha_1, \alpha_2\}$, $\{\alpha_1\}$, and $\{\alpha_2\}$ – α_1 the result of assigning 1 to 'f(a)' and α_2 that of assigning 0 to 'f(a)' – is transitive as well! So, $\{\square f(a), \sim\square\square f(a)\}$, which pans out false under $\langle\{\alpha_1, \alpha_2\}, \alpha_1, R\rangle$, $\langle\{\alpha_1, \alpha_2\}, \alpha_2, R\rangle$, $\langle\{\alpha_1\}, \alpha_1, R\rangle$, and $\langle\{\alpha_2\}, \alpha_2, R\rangle$ for reflexive and transitive R, is sure to do so for reflexive R. I owe the example to McArthur.

(c) The theory $\{\Diamond f(a), \sim\square\Diamond f(a)\}$, though syntactically consistent in $L_3^\square$, is not true under any truth-value T-triple for $L_3^\square$, and this for the same reason as in (b): any reflexive relation on the sets $\{\alpha_1, \alpha_2\}$, $\{\alpha_1\}$, and $\{\alpha_2\}$ in (b) is transitive as well. So, $\{\Diamond f(a), \sim\square\Diamond f(a)\}$, which pans out false under $\langle\{\alpha_1, \alpha_2\}, \alpha_1, R\rangle$, $\langle\{\alpha_1, \alpha_2\}, \alpha_2, R\rangle$, $\langle\{\alpha_1\}, \alpha_1, R\rangle$, and $\langle\{\alpha_2\}, \alpha_2, R\rangle$ for reflexive, symmetrical, and transitive R, is sure to do so for reflexive and symmetrical R. I again owe the example to McArthur.

(d) The theory $\{f(a), \sim\square f(a), \Diamond\square f(a)\}$, though syntactically consistent in $L_4^\square$, is not true under any truth-value T-triple for $L_4^\square$. I owe the example to Weaver.

In all four cases assigning a truth-value (whichever one we please) to one more atomic wff, say 'f(b)', would save the situation. Limiting ourselves to McArthur's first theory, let α_1 be the result of assigning 1 to both 'f(a)' and 'f(b)', α_2 the result of assigning 0 to 'f(a)' and 1 to 'f(b)', and α_3 the result of assigning 1 to 'f(a)' and 0 to 'f(b)'; let α be α_1; and let R be the reflexive – but non-transitive – relation $\{\langle\alpha_1, \alpha_1\rangle, \langle\alpha_2, \alpha_2\rangle, \langle\alpha_3, \alpha_3\rangle, \langle\alpha_1, \alpha_3\rangle, \langle\alpha_3, \alpha_2\rangle\}$. The theory $\{\square f(a), \sim\square\square f(a)\}$ will then be true under $\langle K, \alpha, R\rangle$, as the reader may wish to verify. In other cases truth-values must be assigned to two more atomic wffs, in yet others to three more, etc.

To sum things up, the truth-value of a theory T in a language $L_i^\square$ ($i<5$) is invariant under all possible truth-value assignments to the atomic wffs of $L_i^\square$ that do not figure among the atomic subformulas of T. But, as (a)–(d) show, truth-values must nonetheless be assigned to those wffs if the truth-value of T is to compute right in all cases.

I promised on p. 278 a second – and wider ranging – account of the languages $L_1^\square$–$L_5^\square$. It throws a rather interesting light on the problem at hand.

Call a pair of the sort $\langle \Phi, r \rangle$, where Φ is a function from the atomic wffs of a language $L_i^\square$ ($i=1, 2, 3,$ or 4) to $\{1, 0\}$ and r is a real number, an *indexed function from these wffs to* $\{1, 0\}$, and understand by a *Kripke truth-value triple for* $L_i^\square$ any triple of the sort $\langle K, \alpha, R \rangle$, where K is a non-empty set of indexed functions from the atomic wffs of $L_i^\square$ to $\{1, 0\}$, α is a member of K, and R is a relation on K of the expected sort.[14] This done, take a wff A of $L_i^\square$ to be true under such a triple if conditions (i)–(v) on p. 279 are met, and take a theory T in $L_i^\square$ to be true under the triple if every member of T is.

Following Makinson once more, one can show that:

Theorem 10. A theory T in a language $L_i^\square$ ($i=1, 2, 3,$ or 4) is syntactically consistent in $L_i^\square$ if and only if T is true under at least one Kripke truth-value triple for $L_i^\square$.

Note that T here may be *any* theory in $L_i^\square$, whether infinitely extendible or not. So the account significantly outstrips the one offered on pp. 278–279.[15]

With Kripke triples substituting for the triples on pp. 279–280, proof is again forthcoming that:

(1) the truth-value of a theory T in a language $L_i^\square$ ($i=1, 2, 3,$ or 4) is invariant under all possible assignments of truth-values to the atomic wffs of $L_i^\square$ that do not figure among the atomic subformulas of T,

and proof can *now* be had that:

(2) the truth-value of T will always compute right whether or not truth-values are assigned to the atomic wffs of $L_i^\square$ that do not figure among the atomic subformulas of T.

In particular, let the functions Φ in both $\langle \Phi, 1 \rangle$ and $\langle \Phi, 2 \rangle$ assign the truth-value 1 to 'f(a)', and let the function Φ' in $\langle \Phi', 1 \rangle$ assign it the truth-value 0; let K consist of the three indexed functions $\langle \Phi, 1\rangle$ $(=\alpha_1)$, $\langle \Phi', 1\rangle$ $(=\alpha_2)$, and $\langle \Phi, 2\rangle$ $(=\alpha_3)$; and let α and R be as on p. 281. McArthur's theory $\{\Box \mathbf{f(a)}, \sim \Box \Box \mathbf{f(a)}\}$ will again pan out true under the Kripke triple $\langle K, \alpha, R\rangle$.

So the problem that vexed us on pp. 281–282 has a solution after all: assign truth-values to just the atomic subformulas of your theory T, but outfit the resulting truth-value assignments with indices, thus allowing them to appear under as many different guises as the occasion may call for. Kripke, in effect, does just that when he allows one and the same model to go with different worlds. To the naive his solution sounds of course more exalted, but unlike him I refuse to make metaphysical virtue out of logical necessity. To me indexing one's functions has but one merit: wffs that are irrelevant to the truth-value of a theory can be left truth-valueless, as indeed they should be.

One feature of the above analysis may have puzzled the reader. Only statements are normally said to be true or to be false. Yet many of the atomic wffs – and, on some occasions,[16] all the atomic wffs – whose bearing on the truth-value of a theory I have just investigated are quasi-statements!

So far as first-order languages go, one could skirt the difficulty by banishing all individual parameters and requiring instead that each language L^1 or $L^\Box$ come with aleph$_0$ individual constants. All wffs would now be statements, and theorems in II and III concerning first-order theories would hold true of first-order theories to which infinitely many individual constants are foreign.

Second-order languages would call for similar, though more extensive, tinkering: banish all individual and all predicate parameters, and require instead that each language L^{2_1} of L^2 come with aleph$_0$ individual constants and – for each d from 1 on – aleph$_0$ predicate constants of degree d. All wffs would again be statements, and theorems in II concerning second-order theories would hold true of second-order theories to which infinitely many individual constants and – for any d from 1 on – infinitely many predicate constants of degree d are foreign.

But there is a better answer to the difficulty. Instead of banishing parameters in favor of constants, think of parameters as just so may extra constants!

To illustrate matters, suppose the statements in your theory T come from a language L^1. If T has any model at all, T is sure by Skolem's Theorem to have one whose domain, call it D, is denumerable.[17] So, think of the $\aleph_0$ individual parameters of L^1 as just so many names of the members of D, and the trick is done. Semantically speaking, the quasi-statements of L^1 – and, in particular, those of the sort $\mathbf{F}^d(\mathbf{X}_1, \mathbf{X}_2, ..., \mathbf{X}_d)$ – will now behave exactly like statements, and your assigning them truth-values will be entirely proper.

True, the model that I exploit here, call it M, need not be the *intended* model of T. That model, call it M', might well have a nondenumerable domain D', in which case you will not have names for all the members of D'. But I must urge you to take Skolem's Theorem seriously. Whatever the size of D' in the intended model M', there nonetheless exists at least one model M with a denumerable domain D that suits T as well as M' does, indeed is such that exactly the same statements of L^1 are true in M as are true in M'. So, while others orate in T about M', think M instead, and no one will ever find you out.

Like considerations apply when the statements in T come from a language $L^\square$, a language $L^{2!}$, or a language L^2. Note in the last case that if T has a general model satisfying the (unrestricted) Axiom of Specification, T is sure by Henkin's generalization of Skolem's Theorem to have one whose system of domains, call it $\langle D_0, D_1, D_2, ...\rangle$, consists exclusively of denumerable domains.[18] So, think of the $\aleph_0$ individual parameters of L^2 as just so many names of the members of D_0, of the $\aleph_0$ predicate parameters of L^2 of degree 1 as just so many names of the members of D_1, of the $\aleph_0$ predicate parameters of L^2 of degree 2 as just so many names of the members of D_2, etc., and the trick is done. Semantically speaking, the quasi-statements of L^2 will now behave like statements, and your assigning them truth-values will be above reproach.

Temple University, Philadelphia

NOTES

[1] My thanks go to Brian Chellas, Robert McArthur, and George Weaver for their comments and advice.
[2] Similar results can be had for analogues of my languages $L_1^\square - L_5^\square$ in which neither the Barcan Formula nor its converse are provable. For further details on the matter, see Leblanc (1972b).
[3] The format used here resembles that in Leblanc and Weaver (1972).
[4] Because of the wording of (d) and (g), formulas in which identical quantifiers overlap do not count here as well-formed.
[5] Theories are often presumed to be closed under some provability relation or other. Not so here.
[6] For a short biography of what I call *truth-value semantics*, see Leblanc (1972a).
[7] See in particular Leblanc (1969a) and (1969b).
[8] Theorem 1 holds true, as shown in Leblanc (1969a, b), of any set of wffs of L^i that is infinitely extendible as regards the parameters of L^i, i.e., any set of wffs of L^i to which $aleph_0$ individual parameters and – for each d from 1 on – $aleph_0$ predicate parameters of degree d are foreign. It does not hold of other sets of wffs of L^i, as also noted in Leblanc (1969a). By definition theories are infinitely extendible in this sense; they need not, however, be infinitely extendible in the sense of p. 13.
[9] See Henkin (1950). Note that my concern, when $i=2$, is with *all* general models satisfying the (unrestricted) Axiom of Specification, not just those known as standard models. I shall attend to the latter on a separate occasion.
[10] See Kripke (1963) pp. 86–88 on this matter.
[11] The definition is a generalization by Weaver of one I used in an early draft of this paper.
[12] See Leblanc (1972a).
[13] The predicate constant of degree one 'f' and the individual constant 'a' are presumed to come in (a) from a language $L_1^\square$, in (b) from a language $L_2^\square$, etc.
[14] I call the triples in question Kripke ones to stress their likeness to the models in Kripke (1963).
[15] Proof that Theorem 7 fails for theories which are not infinitely extendible in the sense of p. 13 can be found in Weaver (1972).
[16] I.e., when your language comes without individual constants.
[17] As the reader knows, a model for a language L^1 is a pair $\langle D, I_D \rangle$, where D is a domain and I_D is an interpretation relative to D of the parameters of L^1 (i.e., some result of assigning to each individual parameter and constant of L^1 a member of D and to each predicate parameter and constant of L^1 of degree d ($d=1, 2, 3, ...$) a subset of D^d).
[18] A general model is a pair of the sort $\langle \langle D_0, D_1, D_2, ... \rangle, I_{\langle D_0, D_1, D_2, ... \rangle} \rangle$, where D_0 is a domain, for each d from 1 on D_d is a subset of the power set of D_0^d, and $I_{\langle D_0, D_1, D_2, ... \rangle}$ is the result of assigning to each individual parameter and constant of L^2 a member of D_0 and to each predicate parameter and constant of L^2 of degree d ($d=1, 2, 3, ...$) a subset of D_d. For further details on this whole matter, see Henkin (1950).

BIBLIOGRAPHY

Henkin, L., 1949, 'The Completeness of the First-Order Functional Calculus', *The Journal of Symbolic Logic* **14**, 159–66.

Henkin, L., 1950, 'Completeness in the Theory of Types', *The Journal of Symbolic Logic* **15**, 81–91.
Kripke, S. A., 1959, 'A Completeness Theorem in Modal Logic', *The Journal of Symbolic Logic* **24**, 1–15.
Kripke, S. A., 1963, 'Semantical Considerations in Modal Logic', in *Modal and Many-Valued Logics, Acta Philosophica Fennica* **16**, 83–94, Helsinki.
Leblanc, H., 1969a, 'A Simplified Strong Completeness Proof for QC=,' *Akten des XIV. Internationalen Kongresses für Philosophie* **III**, 83–96, Herder, Vienna.
Leblanc, H., 1969b, 'Three Generalizations of a Theorem of Beth's', *Logique et Analyse* **12**, 205–20.
Leblanc, H., 1972a, 'Semantic Deviations', in *Truth, Syntax, and Modality, Proceedings of the Temple University Conference on Alternative Semantics*, North-Holland, Amsterdam, forthcoming.
Leblanc, H., 1972b, 'On Dispensing with Things and Worlds', in *Existence and Possible Worlds, Studies in Contemporary Philosophy* **2**, New York, forthcoming.
Leblanc, H. and Meyer, R. K., 1970, 'Truth-Value Semantics for the Theory of Types', in *Philosophical Problems in Logic: Some Recent Developments*, D. Reidel, Dordrecht-Holland, pp. 77–101.
Leblanc, H. and Weaver, G., 1972, 'Truth-Functionality and the Ramified Theory of Types', in *Truth, Syntax, and Modality*, forthcoming.
Makinson, D. C., 1966, 'On Some Completeness Theorems in Modal Logic', *Zeitschrift für mathematische Logik und Grundlagen der Mathematik* **12**, 379–84.
Weaver, G., 1972, 'Logical Consequence in Modal Logic: Alternative Semantic Systems for Normal Modal Logics', in *Truth, Syntax, and Modality*, forthcoming.

BRIAN F. CHELLAS

NOTIONS OF RELEVANCE*

Comments on Leblanc's Paper

I. INTRODUCTION

Leblanc entitles his paper 'Matters of Relevance'. He aims to see where and how the truth-value of a set of formulas depends upon the truth-values of the atomic subformulas of the formulas of the set. This is his notion of relevance, here; and he – rightly, I believe – makes no claim that this is the only notion of relevance that matters. More about this in the finale. First, I want to amplify Leblanc's results with respect to modal languages ('languages with a box').

II. SYNTAX

My remarks will solely concern propositional modal languages. The set Fm of (well-formed) *formulas* may thus be regarded as the smallest set (i) including the set Atm = $\{p_0, p_1, p_2, ...\}$ of *atomic formulas*, and (ii) closed under operations of *negation* ($\sim$), *conditionality* ($\supset$), and *necessitation* ($\Box$). Other idioms are defined; in particular: (a) a *truth constant*, **T**; (b) *conjunction*, $\wedge$; (c) *biconditionality*, $\equiv$; and (d) *possibility*, $\Diamond$ (as $\sim \Box \sim$).

A *theory* T is a non-empty set of formulas. By the *atoms* of T, Atm(T), we understand the set of atomic subformulas of formulas in a theory T.

III. AXIOMATICS

By *PC* we mean the set of all (truth-functional) tautologies in Fm. By a *system* we mean a theory including *PC* and closed under *modus ponens*. If S is a system, then by an S-*theorem* we mean a formula in S.

The system $\mathscr{K}$ is the smallest system closed under the rule

$$(R) \frac{A \equiv B}{\Box A \equiv \Box B}.$$

M. Bunge (ed.), Exact Philosophy, 21–27. All Rights Reserved
Copyright © 1973 by D. Reidel Publishing Company, Dordrecht-Holland

By a $\mathcal{K}$-*system* we mean a system including $\mathcal{K}$ and closed under the rule R. The $\mathcal{K}$-systems to be discussed are among those formed by adding to $\mathcal{K}$ various combinations of the following schemata:

K_1. $\Box\mathbf{T}$,
K_2. $\Box(A \wedge B) \equiv (\Box A \wedge \Box B)$,
T. $\Box A \supset A$,
B. $A \supset \Box \Diamond A$,
4. $\Box A \supset \Box \Box A$,
E. $\Diamond A \supset \Box \Diamond A$,
D. $\Box A \supset \Diamond A$.

By K we mean the pair of K_1 and K_2. We denote a system by suffixing to '$\mathcal{K}$' the names of its various characteristic axioms. Thus, e.g., $\mathcal{K}KTE$ is the system got by adding K_1, K_2, T, and E to $\mathcal{K}$ and closing under R (it is better known as Lewis' S5).

In his paper, Leblanc treats of the five systems in the following lattice:

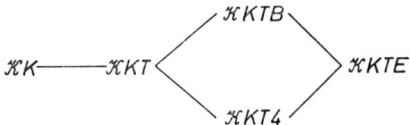

– where $\mathcal{K}K$ is weakest, $\mathcal{K}KTE$ is strongest, and inclusion is marked by the lines. $\mathcal{K}K$, $\mathcal{K}KT$, $\mathcal{K}KTB$, $\mathcal{K}KT4$, and $\mathcal{K}KTE$ are, respectively, Leblanc's $L_1^\Box$, $L_2^\Box$, $L_3^\Box$, $L_4^\Box$, and $L_5^\Box$.

All these systems are extensions of $\mathcal{K}K$ – i.e., all contain K_1 and K_2. We shall be interested, mainly for the sake of a parallel treatment, in the five systems in this lattice:

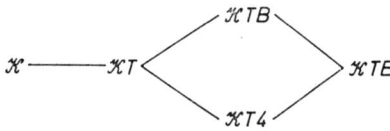

Each system in this latter lattice is strictly weaker than its $\mathcal{K}K$ analogue in the former. (But note that K_1 is a theorem of $\mathcal{K}TB$ and $\mathcal{K}TE$.)

IV. SEMANTICS

$\mathcal{K}$-systems are generally weaker than those with which Leblanc deals, so the sort of semantic theories he provides must be replaced by a more general account.

Where T is a theory, by a $\mathcal{K}$-*model for* T let us understand a structure $\langle W, \alpha, R, \varphi \rangle$ in which W is a non-empty set, $\alpha \in W$, $R \subseteq W \times \mathfrak{P}(W)$, and $\varphi: W \times \text{Atm}(T) \to \{1, 0\}$.

Then we say that a formula A in a theory T is *true in a $\mathcal{K}$-model for* T, $\langle W, \alpha, R, \varphi \rangle$, just in case:

(i)   if $A \in \text{Atm}(T)$, then $\varphi(\alpha, A) = 1$;
(ii)   if $A = {\sim} B$, then B is not true in $\langle W, \alpha, R, \varphi \rangle$;
(iii)   if $A = B \supset C$, then if B is true in $\langle W, \alpha, R, \varphi \rangle$, then C is;
(iv)   if $A = \Box B$, then $\alpha R \{\beta \in W: B \text{ is true in } \langle W, \beta, R, \varphi \rangle\}$.

And if a theory T' is included in a theory T, then T' is true in a $\mathcal{K}$-model for T just in case all its members are.

Now, if C is a class of $\mathcal{K}$-models for a theory T, and A is a formula in T, then we say that A is $\mathcal{K}$ *C-valid in* T just in case A is true in every $\mathcal{K}$-model for T in C. In case C is the class of all $\mathcal{K}$-models for T, we say simply that A is $\mathcal{K}$-valid in T. Also, when T = Fm, A is said to be $\mathcal{K}$ C-valid *simpliciter*.

Next, let W and R be as in a $\mathcal{K}$-model for a theory, and let $\alpha \in W$ and $X, Y \subseteq W$. We note the following conditions:

K'_1. αRW;
K'_2. $\alpha RX \cap Y$ iff αRX and αRY;
T'.   if αRX, then $\alpha \in X$;
B'. $\alpha \in X$ or $\alpha RW - \{\beta \in W: \beta RX\}$;
$4'$.   if αRX, then $\alpha R \{\beta \in W: \beta RX\}$;
E'. αRX or $\alpha RW - \{\beta \in W: \beta RX\}$;
D'.   not both αRX and $\alpha RW - X$.

By K' we mean the conjunction of conditions K'_1 and K'_2. Combinations of the conditions on this list determine classes of $\mathcal{K}$-models for theories, and we agree to denote such classes by juxtaposition of the names of their determining conditions. So, e.g., $K'T'E'$ is the set of $\mathcal{K}$-models for a theory satisfying conditions K'_1, K'_2, T', and E'.

Finally, we may state the following quite general completeness theorem:

Theorem 1. Let $a_1, ..., a_n$ be a selection (possibly empty) from the schemata K_1, K_2, $T, ..., D$ – so that $a'_1, ..., a'_n$ is the corresponding class of $\mathcal{K}$-models (all with respect to a theory, T). Let A be a formula in T. Then:

A is a $\mathcal{K} a_1 ... a_n$-theorem iff A is $\mathcal{K} a'_1 ... a'_n$-valid in T.

When T = Fm above, we obtain as a corollary:

Theorem 2. A is a $\mathcal{K} a_1 ... a_n$-theorem iff A is $\mathcal{K} a'_1 ... a'_n$-valid.[1]

V. TRUTH-VALUE SEMANTICS

So much for the ordinary semantic analysis of $\mathcal{K}$-systems. Leblanc is interested in 'truth-value semantics'. To what extent is this approach adequate?

We wish to imitate Leblanc's constructions, but with a greater degree of generality. So for brevity's sake let us make the following definitions.

Where T is a theory, a *truth-value $\mathcal{K}$-model for* T is a $\mathcal{K}$-model for T in which (i) W is a set of functions from Atm(T) into $\{1, 0\}$, and (ii) φ is absent (or ignored).

A formula A in a theory T is defined to be *true in a truth-value $\mathcal{K}$-model for* T just as in the case of an ordinary $\mathcal{K}$-model for T, except that clause (i) is replaced by:

(i′)   if A ∈ Atm(T), then $\alpha(A) = 1$.

The definitions of the remaining notions – in particular, *truth-value $\mathcal{K}$ C-validity in a theory* T – are defined as in the standard approach, with the prefix 'truth-value' inserted appropriately.

Now when a theory T = Fm, and so Atm(T) = Atm, we have as a limiting case notions of truth-value $\mathcal{K}$-models, truth, and validity *simpliciter*. Let us focus on this case for a moment.

It is easy to show that the various $\mathcal{K}$-systems are complete with respect to the corresponding conceptions of truth-value $\mathcal{K}$-validity. That is, we can prove the following analogue of Theorem 2:

Theorem 3. A is a $\mathcal{K} a_1 ... a_n$-theorem iff A is truth-value $\mathcal{K} a'_1 ... a'_n$-valid.

Furthermore, for a theory T we may define two truth-value $\mathcal{K}$-models (for Fm) to agree on Atm(T) in an obvious way,[2] and reach the result that the truth-value of a theory depends upon the truth-values of its atoms – in the sense that any two truth-value $\mathscr{K}$-models that agree on the theory's atoms agree on all its members. So in this sense we find, like Leblanc, that only a theory's atoms are relevant to its truth-value.

But when a theory $T \neq Fm$ – or at least when $Atm(T) \neq Atm$ – the analogy between the ordinary and the truth-value semantics begins to falter. It is not possible, in general, to give an adequate semantics for $\mathscr{K}$-systems using truth-value $\mathscr{K}$-models *for theories*.

Leblanc has already shown that, except for $\mathscr{K}KTE$, semantics of the sort contemplated does not work for any of the $\mathscr{K}K$-systems he treats. And this goes, too, for other such systems. For example, the formula $\Diamond \Box p_0 \supset \Box \Diamond p_0$ is truth-value $\mathscr{K}K'D'4'$-valid in the theory $\{\Diamond \Box p_0 \supset \Box \Diamond p_0\}$, but it is not a $\mathscr{K}KD4$-theorem.

These negative results extend down to $\mathscr{K}$-systems without K. Consider the following four formulas:

(a) $(\Box \mathbf{T} \wedge \Box p_0 \wedge \Box \sim p_0 \wedge \Box \sim \mathbf{T}) \supset \Box \Box p_0$,
(b) $\Box p_0 \supset \Box \Box p_0$,
(c) $\Diamond p_0 \supset \Box \Diamond p_0$,
(d) $\Box \mathbf{T} \supset (\Diamond \Box p_0 \supset \Box \Diamond p_0)$.

The first, (a), is truth-value $\mathscr{K}$-valid in $\{(a)\}$. The second, (b), is truth-value $\mathscr{K}T'$-valid in $\{(b)\}$. The third, (c), is truth-value $\mathscr{K}T'B'$-valid in $\{(c)\}$. And the fourth, (d), is truth-value $\mathscr{K}T'4'$-valid in $\{(d)\}$. But: (a) is not a $\mathscr{K}$-theorem; (b) is not a $\mathscr{K}T$-theorem; (c) is not a $\mathscr{K}TB$-theorem; and (d) is not a $\mathscr{K}T4$-theorem.

The case of $\mathscr{K}TE$ remains open. I have not been able to discover a counterexample, nor have I been able to prove that truth-value semantics relativized to theories works out all right, as it does with $\mathscr{K}KTE$. (It does work out when a theory has but one atom, because the $\mathscr{K}T'E'$-models for such theories coincide with the $\mathscr{K}K'T'E'$-models. But this does not hold for theories with more than one atom.)

VI. MODELS AND MODALITIES

It is well known that $\mathscr{K}KTE$ has only six genuine (i.e., irreducible) modalities – viz., +(the null modality), $\Box$, $\Box \sim$, and their negations,

∼, ∼□, ∼□∼. The system $\mathcal{K}TE$ also contains (genuinely) just these six modalities. This may help to explain the difficulty in locating a counterexample with respect to truth-value $\mathcal{K}T'E'$-models for theories. It suggests, too, the conjecture that where the number of modalities in a system is finite and small, in some appropriate sense, it is indeed possible to carry out a truth-value semantics as for $\mathcal{K}KTE$.[3]

In this regard, one will wish also to investigate truth-value semantics for the systems $\mathcal{K}KD4E$ and $\mathcal{K}D4E$, since they have the same (genuine) modalities as $\mathcal{K}KTE$ and $\mathcal{K}TE$. I have thus far failed to discover counterexamples for these (and similar) systems, relative to truth-value semantics for theories.

VII. CONCLUSION

In his paper Leblanc seeks to supplant traditional forms of semantic theory with truth-value analyses. I have tried, here, to extend the scope, if not the limits, of his results. But now, in closing, I wish to register some reservations about his notion of relevance.

Leblanc eschews the customary semantic analysis of intensional languages – the so-called 'possible worlds' semantics – as making 'metaphysical virtue out of logical necessity'. And so he would replace such accounts with 'truth-value' analyses. But, alas, these theories are seen to have only limited application. Either all the atoms of a language must be evaluated by the functions in W in a truth-value $\mathcal{K}$-model $\langle W, \alpha, R \rangle$, or the functions must be indexed. Because I share Leblanc's opinion that it ought to be that only the atoms of a theory are *linguistically* pertinent to it, I opt for the second alternative. Here, of course, things work out all right. But, as Leblanc notes, indexing the functions is equivalent to doing semantics in the usual way. In either case, new parameters are required for the evaluation of theories. So more than just the truth-values of a theory's atoms is relevant.

My concern is with natural languages. It has long been recognized – though only more recently well articulated [4] – that an adequate semantic analysis of such languages, with their indexical and intensional features, must relativize the notion of truth to various indices, viz., aspects of the contexts of utterance of sentences. It is clear that the truth-value of 'If your feet hurt, then it may rain tomorrow' depends on more than just the truth-values of its atoms: the speaker, hearer, place, and time of utterance of the sentence are at least to be included among the extra-linguistic features required for an evaluation of the sentence. Such indices thus provide various realizations of the notion of a possible world, and they are all relevant.

University of Pennsylvania, Philadelphia

NOTES

* My thanks to Hugues Leblanc for help in constructing this reply.//
[1] The semantics of $\mathcal{K}$-systems and various completeness proofs are in Cresswell (1970) and Gabbay (1969). Some completeness results covered in Theorems 1 and 2 are new, but not novel. The idea of $\mathcal{K}$-models occurs in Montague (1968, 1970) and Scott (1970). See also Cresswell (1970), p. 349f, n. 6.//
[2] Akin to Leblanc; see following his Theorem 7.//
[3] These remarks reflect, inadequately, a conjecture of Bas van Fraassen.//
[4] E.g., see Bar-Hillel (1954), Davidson (1967, 1970), Montague (1968, 1970), and Scott (1970).

BIBLIOGRAPHY

Bar-Hillel, Yehoshua, 1954, 'Indexical Expressions', *Mind* **63**, 359–79.
Cresswell, M. J., 1970, 'Classical Intensional Logics', *Theoria* **36**, 347–72.
Davidson, Donald, 1967, 'Truth and Meaning', *Synthese* **17**, 304–23.
Davidson, Donald, 1970, 'Semantics for Natural Languages', *Linguaggi nella società e nella tecnica* (ed. by B. Visentini *et al.*), Milan, 177–88.
Gabbay, Dov M., 1969, *Montague Type Semantics for Non-Classical Logics I*, U.S. Air Force Office of Scientific Research, contract No. F 61052-68-C-0036, Scientific Report No. 4.
Montague, Richard, 1968, 'Pragmatics', *Contemporary Philosophy – La Philosophie contemporaine*, Vol. 1, (ed. by Raymond Klibansky), Florence, 102–22.
Montague, Richard, 1970, 'Pragmatics and Intensional Logic', *Synthese* **22**, 68–94.
Scott, Dana, 1970, 'Advice on Modal Logic', *Philosophical Problems in Logic* (ed. by Karel Lambert), D. Reidel Publ. Co., Dordrecht-Holland, pp. 143–73.

PART II

SEMANTICS

LARS SVENONIUS

TRANSLATION AND REDUCTION*

INTRODUCTION

This paper can be considered a commentary on the much-discussed[1] fact that, at least in mathematical contexts, two translations of a theory into another can both be 'correct' without being equivalent in the sense of term-by-term equivalence, or any well-understood sense of equivalence. One popular example of translations which illustrate this fact, is provided by the various translations of number theory into set theory. Much of the discussion deals with this particular example.

The paper is divided into two parts. In the first part the formal concepts are introduced, and some simple purely formal theorems are stated. (Insofar as definitions and theorems could be stated independently of any philosophical theses, I have tried to place them in Part I.) In the second part I try to explain the intended applications of the formal notions which have been introduced, and try to argue for the importance of some of these notions, e.g., the notions 'many-sorted language', 'strict translation', 'weakly definable extension', 'weakly equivalent theory'.

A reader who wants to find out about the main ideas of the paper should probably start with Part II, and occasionally look up a definition in Part I.

PART I. FORMAL PRELIMINARIES

1. *Many-sorted languages*

A *first-order simple many-sorted elementary language* L has the following constituents:

  (i) A non-empty set $\mathscr{S}(L)$ of (primitive) *sort-symbols*. (We will often use $U_1, U_2, U_3, \ldots$ to represent sort-symbols.)

  (ii) For each sort-symbol U, an infinite set of *variables of sort U*.

  (iii) For each sort-symbol U, a (possibly empty) set of *individual constants* of sort U.

  (iv) A set $\mathscr{R}(L)$ of *relation-symbols*, each with an assigned *type*, where a type is a finite sequence $(S_1, ..., S_k)$ of sort-symbols. A relation symbol of type $(S_1, ..., S_k)$ is called a *k-ary* relation symbol.

Among the relation symbols will be, for each type symbol U, a symbol of type (U, U), designated as the *identity symbol of type (U, U)*, and a symbol of type (U), designated as the *universal symbol of type (U)*.

The notion of *formula* in such a language L is defined in the obvious way, using customary connectives and quantifiers. The only feature of the definition which is out of the ordinary is that, when R is a relation symbol of type $(S_1, ..., S_k)$, the only way to use R to make an atomic formula is to put it in a combination $R(t_1, ..., t_k)$, where, for each, i, t_i is a constant or variable of sort S_i.

2. *Interpretations*

When $\mathcal{U}$ is a set of sort-symbols, by a *universe for* $\mathcal{U}$ we mean a function V which to each $U \in \mathcal{U}$ assigns a set $V(U)$ (possibly empty).

An *interpretation J* of a language L (with sorts $\mathcal{U}$) has the followin parts:
 (i) A universe V for $\mathcal{U}$.
 (ii) A function ψ, which
  (a) to each constant c of L of sort U assigns an element of $V(U)$.
  (b) to each relation symbol R of type $(S_1, ..., S_k)$ assigns a subset of $V(S) \times \cdots \times V(S_k)$ (where the identity symbol of type (U, U) is assigned $\{\langle x, x \rangle : x \in V(U)\}$, and the universal predicate of type (U) is assigned $V(U)$.)

The notion of the 'truth-value of a (closed) formula of L under an interpretation J' is defined in the obvious way.

When J_1, J_2 are two interpretations of L (with sorts $\mathcal{U}$), with universes V_1, V_2, and assignments ψ_1, ψ_2, then if Φ is a family of mappings, which to each $U \in \mathcal{U}$ assigns a one-one onto mapping $V_1(U) \to V_2(U)$, we call Φ an *isomorphism* $V_1 \to V_2$.

If such an isomorphism Φ, $V_1 \to V_2$ also 'carries ψ_1 to ψ_2' in the sense that, for each symbol A, $\psi_2(A)$ is in the obvious sense the 'Φ-image of $\psi_1(A)$', we call Φ an *isomorphism* $J_1 \to J_2$.

As should be expected, two isomorphic interpretations give the same truth-values to all closed formulas.

Note: We do not require of an isomorphism $\Phi : V_1 \to V_2$ that it preserve identity or difference between 'objects of different sorts', say between members of $V_1(U_1)$ and $V_1(U_2)$.

A class $\mathcal{M}$ of interpretations of L will be called *regular* if it is closed with respect to isomorphism, i.e., if $\mathcal{M}$ is such that, whenever J_1, J_2 are isomorphic, and $J_1 \in \mathcal{M}$, then $J_2 \in \mathcal{M}$.

When $\mathcal{A}$ is a set of formulas in L, an interpretation J of L is called a *model* of $\mathcal{A}$ if each member of A is true under the interpretation J.

We also write $\mathcal{M}(\mathcal{A})$ for the class of models of $\mathcal{A}$.

The class $\mathcal{M}(\mathcal{A})$ is always regular, in the sense just defined.

3. Theories (in the Formal Sense)

By a *theory* (in the formal sense) we mean a pair $\langle L, \mathcal{M} \rangle$, where L is a language, and $\mathcal{M}$ a regular class of interpretations of L. When $J \in \mathcal{M}$, we call J a *model* of the theory $\langle L, \mathcal{M} \rangle$.

A sentence F in L is called *valid* in $\langle L, \mathcal{M} \rangle$, when F is true under J, for all $J \in \mathcal{M}$.

Note 1. The above definition of 'theory' is wider than the customary definition, since it is not required that $\mathcal{M}$ be the class of models of some set of formulas.

Note 2. In discussing theories in the sense defined, it is sometimes convenient to treat a class of models as if it were an infinite formula, so that we can speak, for instance, of conjunctions, where one conjunct is a class of models, and the other conjunct an ordinary formula. This use should be self-explanatory. – When $\mathcal{M}$ is the class of models of T, we can also call $\mathcal{M}$ 'the axiom of T'.

A language L_1 is called a *sublanguage* of L_2, when all the symbols of L_1 are symbols of L_2.

When L' is a sublanguage of L, J an interpretation of L (with universe V, assignment ψ), then the *L'-reduct of J* is the interpretation J', with universe V' and assignment ψ', determined by the condition that $V'(U) = V(U)$, and $\psi'(A) = \psi(A)$, wherever the left-hand side is defined.

When $T = \langle L, \mathcal{M} \rangle$ is a theory with language L, and L' is a sublanguage of L, then by the *subtheory* $T(L')$ of T (determined by L') we mean the theory $\langle L', \mathcal{M}' \rangle$, where $\mathcal{M}'$ consists of the L'-reducts of members of $\mathcal{M}$.

This is the notion of subtheory which will concern us in this paper.

When $T = \langle L, \mathcal{M} \rangle$ is a theory, L_1 and L_2 two sublanguages of L, and J_1 and J_2 interpretations of L_1 and L_2, resp., we say that J_1 and J_2 are *compatible in* $\langle L, \mathcal{M} \rangle$, when there is a model J of T, such that J_1 and J_2 are the L_1- and L_2-reducts of J.

Two sublanguages L_1, L_2 are called *definitionally equivalent in T*, if for every model of $T(L_1)$ there is exactly one model of $T(L_2)$ which is compatible with the given model in T, and conversely, for every model of $T(L_2)$ there is exactly one model of $T(L_1)$ which is compatible with the given one in T.

When L_1 and L_2 are definitionally equivalent in $T(L_2)$ and L_1 is a sublanguage of L_2, we call $T(L_2)$ a *definitional extension* of $T(L)$.

4. Remark on Defined Sorts

For many purposes it is convenient to extend a language by treating some unary predicates as 'defined sort-symbols', introduce special kinds of variables for these sorts, and introduce defined relations-symbols with types which involve the defined sorts. The term 'simple' in the phrase 'many-sorted simple elementary languages' was used to indicate absence of defined sort-symbols.

Most notions introduced for the case of simple languages, like 'interpretation', 'isomorphism', 'translation', 'correlation', can be extended in a rather obvious way to the case of languages with defined sort-symbols. I have omitted the details, since they are easy to reconstruct.

The main usefulness of this extended notion of language is that it would enable us to simplify important definitions, and to simplify the formulation of theorems.

Instead of giving a systematic treatment of languages with defined sorts, I will here introduce some vocabulary which is inspired by this notion, and which will prove convenient.

Any predicate symbol P of some type (U), (i.e., any unary predicate symbol), will be called a *sort-symbol in the wider sense*. Our original ('primitive') sort-symbols will also be called sort-symbols in the wider sense. When P is a sort-symbol (in wide sense) which is also a symbol of type (U), we say that P is 'subordinate' to U.

When T is a theory, P a non-primitive sort-symbol, and c an individual constant in T, we say that c 'is of sort P in T', if $P(c)$ is valid in T.

When $P_1, ..., P_k$ are sort-symbols in the wider sense, and $U_1, ..., U_k$ are the underlying primitive sorts, and R a relation symbol of type $(U_1, ..., U_k)$, we say that R is *of type* $(P_1, ..., P_k)$ *in* T, if the statement $\forall x_1 ... x_k (R(x_1, ..., x_k) \supset (P_1(x_1) \& ... \& P_k(x_k)))$ is valid in T.

5. Correlations, Correspondences

When S_1 and S_2 are sort-symbols, P, Q, R relation-symbols, with P of type (S_1), Q of type (S_2), R of type (S_1, S_2), then by $\text{Inj}(R, P, Q)$ we mean the formula

$$\forall x_1 \forall x_2 (P(x_1) \,\&\, P(x_2) \supset \exists! y_1 \exists! y_2 (R(x_1, y_1) \,\& $$
$$\&\, R(x_2, y_2) \,\&\, (x_1 \neq x_2 \supset y_1 \neq y_2))$$

and by $\text{Corr}(R, P, Q)$ we mean the formula

$$\text{Inj}(R, P, Q) \,\&\, \forall y (Q(y) \supset \exists x (P(x) \,\&\, R(x, y))$$

We say that R is an *injection* $P \to Q$ in T when the formula $\text{Inj}(R, P, Q)$ is valid in T.

We say that R is a *correlation* $P \to Q$ in T, when the formula $\text{Corr}(R, P, Q)$ is valid in T.

When P is the universal predicate of type U, we also may write $\text{Inj}(R, U, Q)$ instead of $\text{Inj}(R, P, Q)$; and similarly when Q is a universal predicate.

When $\mathcal{U}_1$ and $\mathcal{U}_2$ are sets of sort-symbols (in the extended sense), then any mapping $\sigma: \mathcal{U}_1 \to \mathcal{U}_2$ is called a *sort-mapping* $\mathcal{U}_1 \to \mathcal{U}_2$.

If σ is a sort-mapping $\mathcal{U}_1 \to \mathcal{U}_2$, where $\mathcal{U}_1 = \{U_1, ..., U_k\}$ and C a function defined on $\mathcal{U}_1$, such that $C(U)$ (or 'C_U') for every $U \in \mathcal{U}_1$ is a symbol of type $(U, \sigma(U))$, then we use the notation $\text{Corresp}_\sigma(C)$ for the formula

$$\text{Corr}(C_{U_1}, U_1, \sigma(U_2)) \,\&\, ... \,\&\, \text{Corr}(C_{U_k}, U_k, \sigma(U_k))$$

We call C a σ-*correspondence in* T, if it is a family of symbols of the indicated types, and the formula $\text{Corresp}_\sigma(C)$ is valid in T.

Another convenient notation is the following:

When C is a family of symbols as above (and $\mathcal{U}_1$ is as above), and $t_1, ..., t_k, s_1, ..., s_k$ are terms of sorts $U_1, ..., U_k, \sigma(U_1), ..., \sigma(U_k)$, we use the notation $C \begin{pmatrix} t_1 \dots t_k \\ s_1 \dots s_k \end{pmatrix}$ for the formula

$$C_{U_1}(t_1, s_1) \,\&\, ... \,\&\, C_{U_k}(t_k, s_k).$$

6. Translations

Let T_1 and T_2 be two theories with languages L_1, L_2, $\mathcal{U}_1$ the primitive sorts of T_1, $\mathcal{U}_2$ the sorts (in the extended sense) of T_2, and σ a mapping $\mathcal{U}_1 \to \mathcal{U}_2$.

If ρ is a function which is defined on $\mathcal{U}_1$, and there coincides with σ, and which to the indiv. constants of T_1 correlates individual constants of T_2, and to the relation symbols of T_1 correlates relation symbols of T_2, we call ρ a *σ-translation* $T_1 \to T_2$ if

(1) for any individual constant k in L_1 of sort S, the symbol $\rho(k)$ is of sort $\sigma(S)$.

(2) for any relation symbol Q of T of type $(S_1, ..., S_k)$ the symbol $\rho(Q)$ is of type $(\sigma(S_1), ..., \sigma(S_k))$.

Translations applied to formulas, and to model-classes. When ρ is a σ-translation $T_1 \to T_2$, then when F is any formula of T_1, the *ρ-translation of F'* is the formula obtained by replacing any occurrence of an extralogical symbol S by $\rho(S)$, replacing the variables of any sort U by variables of sort $\sigma(U)$, according to some chosen one-one correspondence, and, in the case of quantifiers $\forall x$ or $\exists x$, where x is of sort S and $\sigma(S)=P$, replacing that quantifier with the 'relativized' quantifier $\forall y_P$ or $\exists y_P$, resp.

When ρ is a σ-translation $T_1 \to T_2$, and J_2 is any model of T_2, one can define in an obvious way an interpretation $\rho_*(J_2)$ of L_1, called the 'inverse ρ-image of J_2'.

When $\mathcal{M}_1$ is any class of interpretations of L_1, θ a translation $T_1 \to T_2$, then by the 'θ-translation of $\mathcal{M}_1$ into T' we mean the class $\mathcal{M}_2$ of models of T_2, defined by $J \in \mathcal{M}_2 \leftrightarrow \theta_*(J) \in \mathcal{M}_1$.

Composition.

When θ_1 and θ_2 are two translations, $\theta_1: T_1 \to T_2$, and $\theta_2: T_2 \to T_3$, there is an obvious way of defining a translation θ_3, which is the 'composition of θ_1 with θ_2' (or $\theta_2 \circ \theta_1$), and such that, for any symbol A of T, $\theta_3(A) = \theta_2(\theta_1(A))$.

Also, if C_1 is a σ_1-correspondence in $T: \mathcal{U}_1 \to \mathcal{U}_2$, and C_2 a σ_2-correspondence $\mathcal{U}_2 \to \mathcal{U}_3$, there is an obvious way of defining the 'composition' $C_2 \circ C_1$, which will be a $\sigma_2 \circ \sigma_1$-correspondence $\mathcal{U}_1 \to \mathcal{U}_3$ (in T).

7. Translation Within a Theory

Let T be a theory, with language L, let L_1 and L_2 be sublanguages of L,

$\mathcal{U}_1$, $\mathcal{U}_2$ sets of sort-symbols (in the wide sense) in L_1 and L_2, resp., and let C be a σ-correspondence $\mathcal{U}_1 \to \mathcal{U}_2$ in T.

Then, when R_1 and R_2 are relation symbols of L_1 and L_2, and R_1 is of type $(S_1, \ldots, S_k)$, we say that R_1 and R_2 *correspond in T via C*, if R_2 is of type $(\sigma(S_1), \ldots, \sigma(S_k))$ and the following formula (called $\mathrm{Csp}(R_1, R_2, C)$) is valid in T.

$$\mathrm{Csp}(R_1, R_2, C): \forall x_1 \ldots x_k x_1' \ldots x_k'$$
$$\left[C\begin{pmatrix} x_1 \ldots x_k \\ x_2' \ldots x_k' \end{pmatrix} \supset (R_1(x_1, \ldots, x_k) \leftrightarrow R_2(x_1', \ldots, x_k')) \right]$$

And, when a is an individual constant of L of sort U, and b a symbol in L of sort $\sigma(U)$, then we use the notation $\mathrm{Csp}(a, b, C)$ for the formula

$$C_U(a, b)$$

And, as before, we say that a and b *correspond in T via C* when $\mathrm{Csp}(a, b, C)$ is valid in T.

When C is a family of symbols of the right type for a σ-correspondence, and θ is a σ-translation $L_1 \to L_2$, we write

$$\mathrm{Support}(C, \theta, \sigma)$$

to refer to the conjunction, whose conjuncts are
  (i) the formula $\mathrm{Corresp}_\sigma(C)$
  (ii) all the formulas $\mathrm{Csp}(E, E', C)$, with E a relation symbol or individual constants in L_1, and $E' = \theta(E)$.

When C is a σ-correspondence in T, and θ a σ-translation, we say that

C supports the translation θ in T

if the formula $\mathrm{Support}(C, \theta)$ is valid in T.

A translation which is supported by a correspondence in T will be called a *strict translation in T*.

A σ-correspondence $C: \mathcal{U}_1 \to \mathcal{U}_2$, such that σ is the identity mapping, and for each $U \in \mathcal{U}_1$, C_U is the identity relation on U, is called the *identity correspondence* $\mathcal{U}_1 \to \mathcal{U}_2$.

A translation supported by an identity correspondence will be called a *trivial translation*.

We call a σ-translation $\theta: L_1 \to L_2$ a 'copying translation' if both the mappings σ and θ are one-one onto, and σ maps the primitive sort-symbols on primitive sort-symbols.

When θ is a copying translation $L_1 \to L_2$, L_1 and L_2 are called 'copies' of each other.

A correspondence C will be called *bijective*, when the underlying sort-mapping σ is one-one onto and maps primitive sort-symbols on primitive sort-symbols.

A translation $\theta: L_1 \to L_2$ (sublanguages of T) will be called *bijective in T*, if it is a copying translation, and is strict in T.

A translation θ (of a language into a sublanguage) will be called *idempotent*, if $\theta \circ \theta = \theta$.

Also a correspondence C is called *idempotent* when $C \circ C = C$.

When T is a theory with language L, L', a sublanguage of L, and θ an idempotent translation $L \to L'$, which is supported by an idempotent correspondence in T, then we call θ a *one-one reduction* of T to the subtheory $T(L')$.

Example: The well-known 'reduction' of arithmetic to set theory should be regarded as a one-one reduction in this sense, of the 'intuitive' combined theory of numbers and sets, to the set-theoretic part of that theory (where this intuitive theory is regarded as a two-sorted theory in our sense.)

8. One-One Extensions of Theories

Let T be a theory with language L, P a 1-ary relation symbol in L of some type (U). Then by a *one-one extension of T grafted on P* we mean a theory formed as follows:

(1) We add one new primitive sort-symbol U' to T, and a new relation symbol R of type (U', U).

(2) We add the axiom $\mathrm{Corr}(R, U', P)$.

The one-one extension of T grafted on P is evidently uniquely determined except for the choice of symbols U' and R.

The definition can be generalized, in an obvious way, to define, for any *set $\mathscr{P}$* of 1-ary relation symbols in L, the *one-one extension of T grafted on the set $\mathscr{P}$ of relation-symbols*.

Prop. When T' is a one-one extension of T formed by adding the sort-symbol U' and the relation symbol R of type (U', U), and T^* is a definitional extension of T', then if C is the family of correlations which assigns R to U', and the identity relation on S to every other sort-symbol

S of T', then C is an idempotent correspondence, and if θ is any idempotent translation $T^* \to T^*$ supported by C, then θ is a one-one reduction of T^* to a definitional extension of T.

(One-one reductions, and one-one extensions, are in a sense inverses of each other.)

9. Weakly Definable Extensions; Weakly Equivalent Theories

Def. When T is a theory, L_1 and L_2 sublanguages of T with $L_1 \subseteq L_2$, then we say that L_2 is a *weakly definable extension* of L_1 in T (and that $T(L_2)$ is a weakly definable extension of $T(L_1)$), if

whenever M_1, M_2, and M_1', M_2' are two compatible pairs of models for L_1 and L_2, and π is an isomorphism $M_1 \to M_1'$, there is exactly one way of extending π to an isomorphism $M_2 \to M_2'$.

Observation: In the case that the (primitive) sorts of the languages L_1 and L_2 are the same, this relation coincides with the relation 'definitional extension' in the classical sense.

Def. Two subtheories T_1, T_2 of T are called *weakly equivalent in T*, if there is some one-one extension T^* of T, such that T^* has some subtheory T_1' which is connected with T_1 by a bijective translation, and such that T_1' and T_2 have a common weakly definable extension in T^*.

Prop. 1. The relation 'weakly equivalent in T' is reflexive, symmetric, and transitive.

Prop. 2. When T_2 is a weakly definable extension of T_1 in T, then T_1 and T_2 are weakly equivalent in T.

Consistent translations.

In the case of distinct theories T_1, T_2 (not subtheories of any bigger theory) we are interested in the following notion.

Def. When $T_1 = \langle L_1, \mathscr{M}_1 \rangle$, $T_2 = \langle L_2, \mathscr{M}_2 \rangle$, a translation $\theta: T_1 \to T_2$ is called a *consistent translation $T_1 \to T_2$*, if there is some common member of the class $\mathscr{M}_1$ and the class $\{\theta_*(J) : J \in \mathscr{M}_2\}$.

(Intuitively, think of T_1 and T_2 as interpreted theories, and consider the problem of 'interpreting' T_1 by giving a translation θ into T_2'. A necessary condition for such an 'interpretation' to be acceptable is clearly that θ is consistent in the above sense.)

For weakly definable extensions we have the following result.

Theorem. Let T_1 and T_2 be distinct theories, with model-classes $\mathscr{M}_1$ and $\mathscr{M}_2$, and let T_1 be a weakly definable extension of T_1'.

Let θ and ρ be two translations $T_1 \to T_2$, such that $\mathcal{M}_2 \subseteq \theta(\mathcal{M}_1)$, (where $\theta(\mathcal{M}_1)$ is the θ-translation of $\mathcal{M}_1$), and $\mathcal{M}_2 \subseteq \rho(\mathcal{M}_1)$. Assume also that θ and ρ coincide on the part T_1'. Then θ and ρ wil be related by strict translation in T_2; i.e. there are strict translations r_1, r_2 in T_2, such that $r_1 \circ \rho = r_1 \circ \theta$.

(To state this roughly, but more intuitively, it says that (under certain conditions) the consistent translation of a weakly definable extension is determined (up to strict translation) by the translation of the subtheory.)

PART II. PHILOSOPHICAL APPLICATIONS

I want here to defend the thesis that many-sorted theories (of the sort described above) are in some ways 'natural' formal representations of many intuitive bodies of knowledge. To establish this thesis I want to argue that we are led to accept many-sorted theories as natural on the basis of insights gained in discussions of 'reductions' of arithmetic to set theory, and similar examples.

On the other hand, there is no claim that a many-sorted representation is ever the only correct representation of an intuitively understood theory. Instead, our theory implies that there is always a great variety of correct representations. This theory of equivalent representations is one useful outcome of the study of many-sorted theories.

I am also arguing that the formal notions 'weakly definable extensions' and 'strict translations', which were introduced in Part I, correspond to important intuitive notions.

1. *Describing the Standard Examples*

We let P be the intuitively understood theory of numbers, formulated in a first-order (one-sorted) language, with extralogical primitive symbols for O, successor, addition and multiplication, as well as for 'number'. We let the symbols be Nu, O, $Succ$, a and p.

We let ST be the intuitively understood theory of sets, also formulated in a first-order language, and with symbols for membership, and also containing symbols for many definable notions which will be specified later. (I could say that I assume we have symbols for *all* definable notions, but it will be obvious that I will not use the full strength of this assumption.)

It is well known that there are various ways of 'interpreting' or 'translating' P into ST (or 'identifying' P with a part of ST). For definiteness, I will mention two such interpretations, the von Neumann interpretation, and the Zermelo interpretation.

The main step in describing each of these interpretations (or 'translations') is to find a set of symbols of ST which will serve as 'translations' of the primitive symbols of P.

We will refer to the von Neumann set of translating symbols as Nu_1, O_1, $Succ_1$, a_1, p_1; and to the Zermelo set as Nu_2, O_2, $Succ_2$, a_2, p_2. The translating functions will be referred to as ρ_1 and ρ_2, resp...[2]

These translations (and lots of others) are agreed to be all equally satisfactory for all mathematical purposes. And it has lately been argued persuasively (by Benacerraf, Quine and others) that there can be no justification (mathematical, philosophical, or other) for singling out one of them and regarding it as 'the correct' translation (as Russell once did).

Although Quine, Benacerraf and others may disagree on certain details, there is some kind of consensus among a substantial group of philosophers (including those mentioned) on what to say about the various translations of P into ST. I will now give some rather vague and preliminary formulations of some parts of this consensus doctrine (to which I subscribe myself.) We will later make use of these theses in some arguments, thereby making them a little more precise.

Some theses about the translations ρ_1 and ρ_2:

(i) Each of these translations is 'correct', or 'preserves meaning'.

(ii) You could teach somebody all there is to 'understand' about arithmetic (in so far as he understands set theory) by giving him one of the standard 'translations' $P \to ST$.

(iii) One thing the discussion has shown is that it is not the 'substance' of what is described by the theory, but only the 'structure', that is of any importance. (This phrase, too vague by itself, might serve to point to 'what is common to the various correct translations'.)

2. *Using Many-Sorted Languages in Analyzing the Translations ρ_1 and ρ_2*

For the sake of the discussion, I will introduce names for some (admittedly vague) philosophical theses:

"*Von Neumann thesis*": The translation ρ_1 is a 'correct' (or 'meaning-preserving') translation $P \to ST$. Or: It is consistent with the meaning of the concepts expressed in P and in ST to make the identifications indicated by the translation ρ_1.

"*Zermelo thesis*": The same statement, except that ρ_2 should be substituted for ρ_1.

Using the notion of many-sorted languages, in our sense, (and the notion of 'embedding theory' to be defined) we will formulate variants of these theses.

Def. When T_1, T_2 are two one-sorted theories, and ρ is a translation $T_1 \to T_2$ (where the universal predicate of T_1 is translated into U^*), then by $\text{Emb}(T_1, T_2, \rho)$ (for 'embedding of T_1 in T_2 by ρ') we mean a theory formed as follows:

(1) We formally adopt different sorts U_1 and U_2 for the universes of T_1 and T_2.

(2) A relation symbol R of type (U_1, U_2) is introduced, with the axioms that R is one-one onto $U_1 \to U^*$, and that R supports the translation ρ.

Remarks: The theory $\text{Emb}(T_1, T_2, \rho)$ is uniquely determined, except for the choice of sort-symbols, and choice of the symbol R. It is obvious how to extend the definition for the case that T_1 or T_2 are many-sorted.

We now introduce the 'weakened theses':

"*Weaker von Neumann thesis*". The theory $\text{Emb}(P, ST, \rho_1)$ is acceptable, in the sense that it can be interpreted consistently with the intended interpretations of P and ST.

"*Weaker Zermelo thesis*". Same statement about $\text{Emb}(P, ST, \rho_2)$.

The stronger von Neumann thesis should be understood to imply the weaker one. Namely: it should follow from the strong thesis that one can interpret the extra relation symbol R in $\text{Emb}(P, ST, \rho_1)$ as identity. The corresponding remark could be made for the two Zermelo theses.

It may be observed that, in the embedding theory $\text{Emb}(P, ST, \rho_1)$ (where R_1 is the symbol for the relation supporting ρ_1), it will also be possible to define a relation R_2, which supports the translation ρ_2. Also, a relation supporting ρ_1 can be defined in $\text{Emb}(P, ST, \rho_2)$. Therefore, the two weaker theses turn out to be equivalent in a simple classical sense.

If there were 'dogmatic von Neumannites' and 'dogmatic Zermeloites', those would still agree that the theory $\text{Emb}(P, ST, \rho_1)$ (with a defined symbol R_2 as above) is correct, but the former would insist that the symbol R_1 must be read as an identity relation, and the latter ones would chose R_2 as an identity relation.

The present-day consensus is that this dispute would be senseless, and that any of these choices is equally faithful to the 'intended' meaning of P and ST.

Still there might be a feeling for some people that it is desirable to make one choice or other rather than leave the theory $\text{Emb}(P, ST, \rho_1)$ as it is. We have got used from classical philosophical semantics to think that each predicate must have a denotation, that two predicates must either have the same or different denotations, and that a predicate must either be an identity predicate or not. The way I want to interpret many-sorted theories (in particular the theory $\text{Emb}(P, ST, \rho_1)$), these matters are sometimes undetermined. And I think a reasonable conclusion from recent discussions is that this should not bother us. Deciding that R_1 is the identity will not add anything essential to our understanding of the theory $\text{Emb}(P, ST, \rho_1)$.

3. A Definability Thesis

We said above (II,1) about the translations ρ_1 and ρ_2, that each seemed to give the correct meaning of the concepts expressed in P (to somebody who did not understand the symbols of P to begin with).

A natural extension of this observation on 'understanding' would be the following:

"If we present to some person A (who understands ST), the axioms of $\text{Emb}(P, ST, \rho_1)$ together with the explanation that the symbols in ST have their usual meaning, this is sufficient to give A full understanding of the symbols in P (and it would not make any difference to his understanding of P whether we told him that P is an identity relation, or left that undetermined.)"

The generalization of this statement to the case of arbitrary theories is the following definability principle:

Definability principle (version 1)

Let T be any many-sorted theory (with one or more sorts); let U be a sort-symbol in T, and P a predicate of type (U).

Then we form an extension T^* of T by

(i) adding a new sort U',
(ii) adding a relation symbol R of type (U', U),
(iii) adding also some new relation symbols involving the new sort U', and possibly some ind. constants of sort U'.

(iv) adding as 'defining axioms',

(A1) The statement that R is one-one onto $U' \to P$.

(A2) For each new relation symbol other than R_1 and each new individual constant, the statement that it corresponds via R to some specified relation symbol of the corresponding U'-less type.

This extension T^* will then be called a 'definitional extension' of T; and our principle says that the meaning of the new symbols of T^* is completely determined by the so-called defining axioms, and the meaning of T.

(This formulation of the principle can be simplified. We will return to this matter later.)

This principle gives us a licence to expand a given theory, adding new sorts, and specifying the meanings of the new symbols by means of one-one mappings as indicated (in the spirit, in which classical-type definitions are introduced). It should be observed, that by classical standards, the meaning of the new symbols in the extension T^* is not completely determined. My thesis is, that if we take seriously the observations concerning the 'indifference' between ρ_1 and ρ_2, we should admit that one-one mappings of this kind do determine meaning. One could regard this definability principle as an application of the vague thesis stated in II, 1, that 'only the structure is important'.

Some philosophers feel that this indifference doctrine which I have argued for, and which is the basis of our new definability principle, may be correct when applied to mathematical theories, but not correct when applied to empirical theories. My feeling is that there is no reason to draw this limit for the application of the theory. (At least there should be some arguments for such a limit, and I have not seen any.) Quine (*Word and Object*) has some arguments pointing in the direction that two adequate explanations of meaning (of an empirical term) need not be equivalent in any classical sense, and that the situation in an empirical subject is rather like the situation with the interpretation of arithmetic.

I will point out some consequences of this definability principle.

Example 1. We consider again some version of ST, with sort-symbol U_S, and the usual von Neumann arithmetical symbols Nu_1, O_1, $Succ_1$, a_1, p_1.

There will be in ST some non-trivial relation-symbol A, for which it is provable in ST that A is a one-one onto mapping $Nu_1 \to Nu_1$.

Let us form a two-sorted extension ST^* of ST as follows:

(i) We add a new sort-symbol U_P, two relation symbols R, R' of type (U_P, U_S), and symbols $O, Succ, a, p$, and $O', Succ', a', p'$ (of obvious types involving the new sort)

(ii) We add the axioms

(Ax 1) R and R' are one-one onto $U' \to Nu_1$

(Ax 2) $R(x, u) \, \& \, A(u, v) \supset R'(x, v)$

(Ax 3) The statement that $O, Succ, a, p$ correspond to $O_1, Succ_1, a_2, p_1$, via R.

(Ax 4) The statement that $O', Succ', a', p'$ correspond to $O_1, Succ_1, a_2, p_1$ via R'.

We could establish, by some simple computations, that this theory is a definitional extension, in a classical sense, of the part which only involves the new symbols $R, O, Succ, a, p$. Since this part is a definitional extension of ST by our principle, the bigger theory ST^* is also a definitional extension of ST in our sense.

By our principle, then, the meaning of the new symbols is completely determined by the axiom of ST^* cnd our understanding of ST. But intuitively it seems that we have a choice whether to understand the set $\{O, succ, a, p\}$ or the set $\{O', Succ', a', p'\}$ as representing the standard arithmetical notions. If somebody, in an effort to make the meaning more definite, says; I take O to mean zero, $Succ$ to mean 'successor' etc. then by our principles he has not added anything regarding the meaning. We have then shown that, according to our principles, one can 'completely understand' a theory without knowing which symbol represents O (etc.)

(We will be able to state this observation in a different terminology in the discussion of 'correct translations')

Example 2. In discussing formulas and other expressions of formal languages of set theory or other theories, it is generally recognized that the expressions 'could be regarded' as numbers, and the different ways of combining symbols and formulas (e.g. juxtaposition) could be regarded as arithmetic functions.

Formally, the situation here is very much like the situation with arithmetic and set theory. And we could consider the same moves, of 'defining', in our extended sense, some predicates Fla, Vble, Appl etc., of a new sort. To make the problems more explicit we could even (as we did in the first example) simultaneously introduce several sets of new symbols like that, corresponding to different definable mapping relations.

According to our definability thesis, these axioms should completely determine the meanings of the introduced symbols. And if somebody adds: This symbol is to be understood as 'variable', etc. we should say that he has not added anything essential to our understanding. – It seems that in this case our emotional resistance might be greater (we think of formulas as something that 'could be written down'), but I would still urge that this consequence be accepted.

One could consider other examples: where the roles of 'lines' and 'surfaces' are interchanged, for instance. The kinds of objections we feel here are: "But surfaces and lines *are* quite different, so it makes a real difference in my understanding of the language, whether I read this symbol as 'surface', and that symbol as 'line', or vice versa." But there is also a real difference between even numbers, and odd numbers; and I would suggest that the cases could be treated in the same way.

4. *Additional Remarks on the Definability Principle*

The formulation of the definability principle given in Section 3 was chosen because it is close to the interesting applications of the principle.

However, the principle can be formulated equivalently (equivalently, that is, via classical definability principles) in the following simpler way:

Definability principle (version 2).

Let T be any theory, U one of its sort-symbols, and P a predicate of type (U).

We form an extension T^* of T by

(1) adding a new type-symbol U', and a relation symbol R of type (U', U).

(2) adding as a new axiom ('defining axiom') the statement that R is one-one onto $U' \to P$.

Our principle then states, that the meaning of the new symbol R is completely determined by the defining axiom and the meaning of T.

(In this formulation (version 2) we consider a much simpler extension of T than in version 1. However, the extension considered in version 1 is just a classical definitional extension of this one.)

We do want to accept the generalization of this principle to the case where many new sorts are introduced, via one-one mappings. Making use of the fact that one can introduce 'defined' sort-symbols in T, this generalization can be stated most simply as follows:

Definability principle for many new sorts:

When T^* is a one-one extension of T (I, 8), then the meaning of the new symbols of T^* is completely determined by the meaning of the symbols in T, and the axioms of T^*.

I have tried to argue that the insights gained in considering various 'interpretations' of arithmetic into set theory forced us to accept our present definability principles.

The question arises whether these principles can be strengthened in some natural way; or, whether the same sorts of reasons which support the principles we have formulated also would support something stronger. I propose that this is the case for the following:

Strengthened definability principle.

When T^* is any weakly definable extension of T(I, 9) then the meaning of the new symbols of T^* is completely determined by the meaning of T, and the axioms of T^*.

I have a justification for this principle which I think has some force. For lack of space I will not try to develop this argument here, but I may mention that it is a rather direct application of the theorem on weakly definable extensions which is stated in I, 9.

I want to mention one special type of examples of weakly definable extensions. (called 'sort-unifying extensions')

Def. Let T be a theory, U_1 and U_2 two distinct sorts in T.

The 'extension T^* of T which unifies U_1 and U_2' is formed as follows:

(1) We add a new sort-symbol U', a relation symbol R_1 of type (U_1, U'), and a relation symbol R_2 of type (U_2, U').

(2) We add as 'defining axioms'

(a) Statements saying that R_1 and R_2 are one-one mappings into

(b) The statement

$$\forall y (\exists x_1 R(x_1, y) \leftrightarrow \sim \exists x_2 R(x_2, y))$$

It can be verified that an extension formed in this way is a weakly definable extension.

5. Correct Translations

We saw before that when a theory is formed as a one-one extension, we actually provide a 'translation' for the introduced symbols into old symbols (and, in fact, for all the formulas of the extended theory into formulas of the old theory) which should be regarded as 'meaning-preserving', or 'correct'. This translation is now always a strict translation, in the sense we have defined. Once we accept that these special strict translations should be regarded as correct, there is an easy argument to convince us that all strict translations between subtheories of T should be regarded as correct. (According to my view, there might be correct translations in T other than the strict translations, but strictness is the only formal criterion that we have.)

We will also assume that, in the case of two distinct languages (thought of as languages of two persons, or of two cultures), one can sometimes say that a part of one means the same as a part of the other, or even that the one theory can be 'identified' with some part of the other. Using another terminology, this means that we assume that there are sometimes correct translations from a part of one language (or theory) to a part of another language (or theory).

One requirement for a translation $\theta: T_1 \to T_2$ to be called correct is evidently that it be *consistent* in the sense defined in I, 9.

We can strengthen this to:

A translation $\theta: T_1 \to T_2$ is correct, just in case the embedding theory $\text{Emb}(T_1, T_2, \theta)$ is correct.[3]

We also want to assume about the correct translations that they are closed under composition; in particular, we assume, that when θ is a correct translation $T_1 \to T_2$, and ρ_1 and ρ_2 are correct translations $T_1 \to T_1$ and $T_2 \to T_2$, resp., then $\rho_2 \circ \theta \circ \rho_1$ is a correct translation $T_1 \to T_2$.

It follows from this, that a correct translation $T_1 \to T_2$ is not necessarily one-one, and does not necessarily preserve difference of sorts.

We return now to a previous example.

We discussed in II, 3 a theory which contains two different 'acceptable' sets of notions for arithmetic, and we argued that it must be undetermined which set is taken as representing the standard arithmetical notions. In terms of correct translations, this claim could be restated as follows (thinking of a correct translation $T_1 \to T_2$ as a way of 'giving the meaning of T_1').

Let the theory ST^* considered in the example be the language of a person A, and let B be a person with a theory T, which contains a language of arithmetic (with symbols Nu_B, O_B, $Succ_B$, etc., with the usual meanings). Then there will be a correct translation θ of the arithmetical part of ST^* into T, such that $\theta(O) = O_B$, $\theta(Succ) = Succ_B$, etc. But there will also (according to our principles concerning correct translations) be a correct translation θ', such that $\theta'(O') = O_B$, $\theta'(Succ') = Succ_B$, etc.

The remarks just made, on the possibility of 'translating' a part of ST^* in different ways, could be taken as a way to interpret the previous remarks in II, 3 in the discussion of example 1, on the possibility of 'choosing' symbols for arithmetical notions in different ways.

A theorem on intertranslatable theories.

Assume that T_1 and T_2 (with languages L_1 and L_2) are two theories which are 'intertranslatable' in the sense that there are correct translations $\rho_2: T_1 \to T_2$ and $\rho_2: T_2 \to T_1$.

Then there are correct strengthenings T_1^* and T_2^* of T_1 and T_2, with the same languages L_1, L_2, which have a common one-one extension.

Proof: By our listed requirements for correct translations, the assumption that ρ_1 and ρ_2 are both correct means that a combined theory of $\mathrm{Emb}(T_1, T_2, \rho_1)$ and $\mathrm{Emb}(T_2, T_1, \rho_2)$ is correct (with different symbols chosen for the introduced correspondences in each case.)

When we call this combined theory T^*, we can verify that T^* is a one-one extension of each of the reduct-theories $T^*(L_1)$ and $T^*(L_2)$. $T^*(L_1)$ and $T^*(L_2)$ are the correct strengthenings of T_1 and T_2 which are required by the theorem.

6. Simple Examples of Weakly Equivalent Theories

(1) Let T_1 be a theory, with just one sort U_1 one 1-ary predicate A_1, and just the models with the structure

$$\bar{U}_1 = \{a, b\}, \qquad \bar{A}_1 = \{a\}$$

(2) We could form T_2 by 1-1 extension from T_1 as follows: Sorts U_1, U_2, relation symbol R of type (U_2, U_1); A_1 as before, A_2 of type (U_2), and models with the structure

$$\bar{U}_1 = \{a, b\}, \qquad \bar{U}_2 = \{c, d\}, \qquad \bar{R} = \{\langle a, c\rangle, \langle b, d\rangle\},$$
$$\bar{A}_1 = \{a\}, \qquad \bar{A}_2 = \{c\}.$$

(3) By adding to T_2 a *unifying sort* U_3, with injection relations Q_1 and Q_2, we get a theory T_3 with symbols
  (1) Those of T_2
  (2) Also U_3, Q_1, Q_2, and B_1, B_2, V_1, V_2 (of type (U_3)).
The 'defining' axioms will be

(Axiom 1) The standard axioms saying that U_3 is a unifying sort via the relations Q_1 and Q_2.

(Axiom 2) (a) $B_1(z) \leftrightarrow \exists x (Q_1(z, x) \,\&\, A_1(x))$
(b) $B_2(z) \leftrightarrow \exists y (Q_2(z, y) \,\&\, A_2(y))$
(c) $V_1(z) \leftrightarrow \exists x Q_1(z, x)$
(d) $V_2(z) \leftrightarrow \exists y Q_2(z, y)$

(T_3 is what we have called a 'sort-unifying extension' of T_2)

(4) We let T_4 be the subtheory of T_3 formed by dropping the sorts U_1 and U_2. The vocabulary will then be:

$$U_3, V_1, V_2, B_1, B_2.$$

(We can see that T_3 is a one-one extension of its subtheory T_4. The relation between T_4 and T_2 is also of some interest. We might say that 'T_4 is the theory obtained from T_2 by unifying the sorts').

According to our definitions, all the theories T_1–T_4 would be weakly equivalent; and according to our thesis, they all could be used to 'say' the same things, or to convey the same information.

University of Maryland

NOTES

* This work was supported by a grant from the General Research Board, University of Maryland.
[1] See, for instance, Benacerraf (1965) and Quine (1969).
[2] For details on these translations, see for instance Quine (1963), §§ 12–16.
[3] The notion of 'correct theory' used here cannot, it seems, be understood in the sense of 'true theory', but rather something like 'theory which there are good reasons to accept'. This means that 'correct theory' is a kind of modal notion. It will follow that 'correct translation' is also a modal notion.

BIBLIOGRAPHY

Benacerraf, Paul, 1965, 'What Numbers Cannot Be', *Phil. Rev.* **74**.
Quine, W. V., 1960, *Word and Object*, Ch. II, MIT Press, Cambridge, Mass.; John Wiley, New York.
Quine, W. V., 1963, *Set Theory and its Logic*, The Belknap Press of Harvard University Press, Cambridge, Mass.
Quine, W. V., 1969, 'Ontological Relativity', in *Ontological Relativity and Other Essays*, Columbia University Press, New York.

MARIO BUNGE

A PROGRAM FOR THE SEMANTICS OF SCIENCE

I. PROBLEM, METHOD AND GOAL

So far exact semantics has been successful only in relation to logic and mathematics. It has had little if anything to say about factual or empirical science. Indeed, no semantical theory supplies an exact and adequate elucidation and systematization of the intuitive notions of factual reference and factual representation, or of factual sense and partial truth of fact, which are peculiar to factual science and therefore central to its philosophy. The semantics of first order logic and the semantics of mathematics (i.e., model theory) do not handle those semantical notions, for they are not interested in external reference and in partial satisfaction. On the other hand factual science is not concerned with interpreting a theory in terms of another theory but in interpreting a theory by reference to things in the real world and their properties.

Surely there have been attempts to tackle the semantic peculiarities of factual science. However, the results are rather poor. We have either vigorous intuitions that remain half-baked and scattered, or rigorous formalisms that are irrelevant to real science. The failure to pass from intuition to theory suggests that semanticists have not dealt with genuine factual science but with some oversimplified images of it, such as the view that a scientific theory is just a special case of set theory, so that model theory accounts for factual meaning and for truth of fact. If we wish to do justice to the semantic peculiarities of factual science we must not attempt to force it into any preconceived Procrustean bed: we must proceed from within science. We should realize that a scientific theory is more than its mathematical formalism, and that this surplus is not describable in terms of 'operational definitions', let alone 'ostensive definitions'

My proposal is to explore and implement the following program for the semantics of science:

Problem: To investigate the semantic aspects of scientific theories.

Method: (i) To start by analyzing real specimens of scientific theory with a view to disclosing its semantic components – mainly reference, representation, meaning, and degree of factual truth. (ii) To build exact (i.e., mathematical) theories about these semantic notions and their cognates. (iii) To check whether the explicata thus obtained are adequate, or at least relevant to live science.

Goal: To articulate the various special theories into a semantic theory of science capable of performing the following jobs. (i) To clarify and systematize the semantic aspects of scientific theories as distinct from the semantics of formal theories. (ii) To help scientists determine the precise reference and sense of their own theories – which reference and sense are often the object of heated debate.

This paper will outline the principles chosen to implement this program and will report briefly on some of the results obtained so far.

II. GUIDELINES

The semantical theories we wish to build should spell out the following principles:

(i) The symbols in a conceptual language designate constructs (concepts, propositions, or theories). In particular, a sentence is one among a number of signs designating a proposition. In short, we espouse conceptualism rather than literalism – not however a conceptualism of the Platonic variety.

(ii) Some of the constructs employed in science refer to real or supposedly real objects, such as protons, dinosaurs, and tribes. The set of putative referents of a factual construct may be called the latter's reference class. In other words, we adopt (critical) realism rather than either conventionalism or any form of subjectivism (e.g., operationism).

(iii) The reference class of a factual proposition and of its negate are the same. (No negative facts.) And the reference class of a truth functional compount of two or more propositions equals the union of the reference classes of the components.

(iv) Some factual constructs represent certain traits of their referents. For example, the atomic number of an atom represents the number of protons in its nucleus; and the matrix of the probabilities that the individuals in a community make transitions from one social layer to the other strata, represents the social mobility of the community. Such representations are literal not metaphoric, and symbolic rather than iconic. A factual theory, when formulated explicitly, should include statements indicating what the referents of its basic concepts are and what if anything they represent.

(v) Factual constructs have both an external reference and a sense. Two predicates representing different properties of one and the same thing differ in sense. For example, the concepts of electric conductivity and thermal conductivity are coreferential and even coextensive but not cointensive. Consequently we need a nonreferential theory of sense. Sense and reference are not mutually reducible: they must be treated on a par. They constitute the two components of meaning. Metaphor: regard R and S as the radius vector and polar angle, respectively, of a vector (meaning) in the plane of constructs. All the constructs with the same reference class (e.g., the various thermodynamic functions of a piece of matter) are represented by vectors with tips lying on a common circle. All the constructs with the same sense but different referents (e.g., the temperature values of different bodies) are collinear. A change in both reference and sense is represented by a pair of vectors which are neither collinear nor on a common circle.

(vi) The sense of a representing factual construct, such as a position coordinate or a mutation rate, is given by (a) its mathematical structure and (b) that which it represents. In an axiomatized factual theory both aspects should be taken care of (not just the formal aspects). And in such a theory it is the axioms in which the construct occurs that ultimately determine its sense – or, as we may say, such axioms determine the gist of the construct.

(vii) Sense is contextual: strictly speaking there are no categorematic terms. While extrasystematic sentences are hardly significant, the sense of a construct belonging to a theory is assigned by a good portion of the whole theory – in fact by all the constructs that are logically related to the given construct. Change the theory and 'the same' construct (or rather the same symbol) is likely to change its sense, even though its reference may remain invariant. Therefore the concept of sense should be relativized to a theory.

(viii) In a scientific theory sense and reference are either assumed or derived. The search for sense must therefore proceed both upwards, to the basic assumptions, and downwards, to their logical consequences. The former will constitute the gist, the latter the content of the construct. A deductively isolated predicate, if there were any, would have no precise sense at all.

(ix) Conjunction enriches. Therefore the sense of a conjunction should include the senses of the conjuncts. The sense of a negation should equal the complement of the original sense in the given context. And if two constructs are identical so must be their senses. These assumptions should suffice as a foundation for a theory of sense.

(x) The 'inverse law' of intension and extension should hold when formulated in this way: "If the sense of A is included in the sense of B, then the extension of A includes (or is equal to) the extension of B". This formula should be a theorem in the theory of sense.

(xi) Any talk of meaning variance or invariance should be accompanied by a theory of meaning – otherwise it will be just loose talk. The 'amount' of change in the meaning (sense *cum* reference) of a construct when adopted by a new theory should be expressible in exact (e.g., set theoretic) terms.

(xii) From a semantic standpoint a factual theory is an interpretation of a mathematical formalism. One and the same formalism may be assigned alternative factual interpretations, each of which gives rise to a different factual theory. A factual interpretation is, roughly, an assignment of factual meaning. In other words, a construct in factual science is a mathematical construct together with a factual interpretation.

(xiii) Whereas interpretation bears on exact concepts, elucidation bears on inexact ones. Interpretation is, roughly, the converse of elucidation. Thus probability elucidates or exactifies the concept of possibility and, conversely, possibility interprets probability. (Incidentally, for this reason, i.e., because there exists a quantitative calculus of possibility, science makes no use of modal logics.)

(xiv) Meaning is prior to truth – *pace* Frege and the Vienna Circle. Change the interpretation of a formula and its truth value may change. Moreover, most formulas are never tested for truth, hence are never assigned a truth value: they have to wait in a semantic limbo. And yet they are supposed to satisfy the laws of logic and to have a definite meaning. And such a meaning must be understood before any experiment can be designed to find out truth values.

(xv) Truth conditions, important as they are in elementary logic, become blurred in science. To begin with, a truth condition for a factual sentence cannot possibly determine the significance of the latter. (While every statement comes with a more or less definite meaning, it may not have been assigned a truth value. And acquiring one won't change its sense and reference.) Factual truth conditions are the business of scientists and methodologists, not of semanticists. And, rather than clear cut biconditionals ("⌜A⌝ is true iff B"), in actual practice a truth condition consists of an ill articulated set of necessary conditions for high, medium, or low degree of factual truth. Moreover, the assignments of factual truth values are provisional.

(xvi) Truth, a semantic property, is a property of propositions not of physical objects such as written or spoken sentences. And factual truth is a property of statements with a factual reference. But complete truth is not easy to come by in factual science: the best we get is approximate truth. Moreover we always get relative truth, i.e., truth degree gauged against some proposition taken to be true. For example, let F and G be two functional statements representing, each in its own theory, a given feature of a supposedly real thing, such as the velocity of a body falling freely in the vacuum. In particular, suppose that

$$F = \ulcorner v = v_0 \urcorner \quad \text{(Aristotle)}$$
$$G = \ulcorner v = v_0 + gt \urcorner \quad \text{(Galilei)}.$$

A possible formula for the truth value of F given G (assuming G to be true) is the absolute value of the ratio of the values of the functions for the same thing:

$$V(F \mid G) = \frac{v_0}{v_0 + gt}.$$

This relative degree of truth approaches 0 (is near unity) for large (very small) values of g or of t. Alternative formulas are possible.

This being the case it behoves the semanticist to elucidate this concept of relative and approximate truth and, in general, the concept of degree of truth. Since this concept of partial factual truth is important, it is unlikely to be definable. The best strategy may be to make it into a primitive concept of a special theory.

(xvii) In attempting to build a theory of partial and factual truth we must resist the temptation to equate it with probability or some function of probability. The main reason for this is that there seems to be no procedure for assigning probabilities to propositions other than by arbitrary fiat. We must also resist the temptation to resort to many-valued logics. The main reason for this is that mathematics, the skeleton of factual science, has ordinary logic built into it. A theory of partial and relative truth of fact should then presuppose ordinary logic. To this end, the logical truth values may be regarded as just the unit and the zero elements of the Boolean algebra of (formally equivalent) statements. This structure can then be imposed any number of alternative metrics. Any member of the unit element of the set (i.e. Frege's *das Wahre*) can be assigned the real number 1, and any member of the zero element of the set (*das Falsche*) the real number 0.

(xviii) Desiderata for a theory of partial truth consistent with ordinary logic: (a) Truth is a (partial) function from the set of propositions into an interval of the real line, e.g., [0, 1]; i.e., $V(p) = v \in [0, 1]$. (b) $V(\sim p) = 1 - V(p)$. (c) If the propositions p and q are logically independent (not interdeducible), then $V(p \& q) = V(p) \cdot V(q)$.

(xix) The notion of extension, though important, is derivative, for it depends on the concepts of reference and of truth. The strict extension of a concept is the collection of those of its referents that happen to have the property represented by the concept. The lax extension of a concept (whether exact or vague) is the set of referents that satisfy it approximately or to a given extent. (Incidentally, do not mistake 'extensional' for 'truth-functional' and 'intensional' for 'non-truth functional', as *PM* did. Science is not purely extensional, as every one of its constructs comes with an intension. But science employs only ordinary (truth functional) logic. If for no other reason the semantics of science has no use for what are often, mistakenly, called "intensional [non truth functional] logics".)

(xx) Lastly, a piece of methodological advice: In expanding the preceding principles into theories, try and keep them together. Do not attempt to handle each semantic concept in isolation but try to articulate the theories of the various concepts (reference, representation, sense, truth, extension, etc.). The reason is plain: these concepts *are* inter-related.

III. PREVIEW OF RESULTS

Here go, in quick succession, some inklings of the results obtained so far in implementing the program formulated in the previous section.

(i) *Designation* is construed as a certain many-one function from signs to constructs.

(ii) *Reference* is elucidated as a certain function from constructs to things. More exactly, two reference functions are introduced, one from predicates to sets of individuals, the other from propositions to sets of individuals. And the factual reference functions are the restrictions of the preceding functions to sets of factual items.

(iii) *Denotation* is defined as the composition of designation and reference.

(iv) *Representation* is clarified as a certain relation from constructs to aspects of things. Whatever represents refers but not conversely.

(v) Two representing constructs in a theory constitute *equivalent representations* of the same factual item if they can be freely substituted for one another in every basic law statement of the theory.

(vi) The *purport* or upward sense of a construct x in a set C of constructs, closed under the logical operations, is the principal ideal generated by x in C. In other words, the purport of a construct is the collection of its logical forebears.

(vii) The *gist* or essential sense of a construct is a subset of its purport. In an axiomatic theory the gist of a construct is the set of axioms in which the construct occurs. Whence the semantic import of axiomatics.

(viii) The *import* or downward sense (or content) of a construct x in a context C closed under the logical operations is the principal filter generated by x in C. That is, the import of a construct is the totality of its logical progeny.

(ix) The *full sense* of a construct is the union of its principal ideal and its principal filter, i.e., of its purport and import.

(x) If a construct has no place in a deductive system, or if its place is ignored, we assign it a horizontal sense or *intension* (or comprehension). If p and q are constructs of the same type (either predicates or propositions and if they can be conjoined, then (a) $I(p \& q) = I(p) \cup I(q)$; (b) $I(\neg p) = \overline{I(p)}$; (c) if $p = q$ then $I(p) = I(q)$ but not conversely.

(xi) Consider a Boolean algebra of either predicates or propositions. Then the family of their intensions is a ring I of sets: the ring of intensions. Define in I the function $\delta : I^2 \to I$ with $\delta(p, q) =$ The symmetric difference between the intension of p and the intension of q. Then δ defines neighborhoods in a topological space. And a neighborhood of p in this space is constituted by the conceptual relatives of p. This elucidates Wittgenstein's vague notion of *family resemblance*.

(xii) *Meaning* is taken to be a property of constructs. The meaning of x in C is the ordered pair: ⟨Sense of x in C, reference class of x in C⟩. Consequently two constructs have the same meaning just in case they have both the same sense and the same referents, i.e., if 'they' are the same construct.

(xiii) *Signification* is regarded as a property of signs. It may be construed as the composition of designation and meaning. The significance of a sign is the meaning of the construct the sign designates. Signs may thus have a vicarious meaning (sense *cum* reference). A sign is nonsignificant just in case it designates no meaningful construct.

(xiv) *Synonymy* (in a language) is equal significance, hence identical meaning (same sense and reference) of the underlying constructs. We can do better than just defining synonymy: we can compare symbols as to significance, since our theory of significance rests on a theory of meaning couched in set theoretic terms.

(xv) *Truth* is construed as a real function V on a subset S_D of the set S of statements. (Being a partial function on S, V makes room for truth value gaps, which are only too conspicuous in science.) Because the family of equivalence classes of the propositions in S_D has a Boolean structure, the whole thing becomes a metric Boolean algebra.

(xvi) The desiderata xviii imposed on V in Section II determine a function V that looks adequate, in the sense that it seems to be consonant with actual patterns of scientific inference. Moreover, actual statements in theoretical science can in principle be assigned definite *degrees of truth* (but not probabilities). Thus if a statement p has been found by experiment to be in error by the amount ε, we set $V(p) = 1 - \varepsilon$, and we deduce $V(\sim p) = \varepsilon$. In this way the theory of truth can be conjoined with the theory of scientific inference (a branch of mathematical statistics, not of inductive logic).

(xvii) The quantitative concept of truth allows us to define certain qualitative concepts such as the one of set of confirmers (or else of infirmers) of a given statement. In fact, consider the (quasi) distance function $d: S_D \times S_D \to [0, 1]$ such that $d(p, q) = |V(p) - V(q)|$ for any p and q in S_D. An open neighborhood of a point p (the family of its alethic relatives) is the *set of confirmers* of p.

(xviii) *Definite description* loses much of its glamour in our semantics, for it is construed as indicating uniqueness rather than both uniqueness and existence. Moreover, in one of our construals 'The length of a' is just the first half of the complete functional statement 'The length of a equals b'.

(xix) *Analyticity*, central to the semantics of logic and mathematics, is rather unimportant in the semantics of science provided it is conceived in a narrow way. The construal I propose is this: A formula is analytic in a given theory iff it holds under all interpretations (in all models) of the theory or is a definition in the theory. The great divide is not analytic/synthetic but formal/factual. And the analytic formulas constitute a smallish (though infinite) subset of the set of formal formulas.

(xx) The upshot of our investigation will be a body of theories that may be regarded as included in, or at least tangential to, epistemology. Indeed, they are concerned with constructs belonging to the body of our conjectures about the world. The relation of this semantical theory to metaphysics is but slight. We distinguish three concepts of existence: neutral [paradigm: $(\exists x) Px$], conceptual [paradigm: $(\exists x)(Px \ \& \ x$ is a construct)], and physical (paradigm: $(\exists x)(Px \ \& \ x$ is a physical object)]. Existential quantification, unless qualified, is ontologically neutral. Logic and mathematics have nothing to do with ontology except that they should be

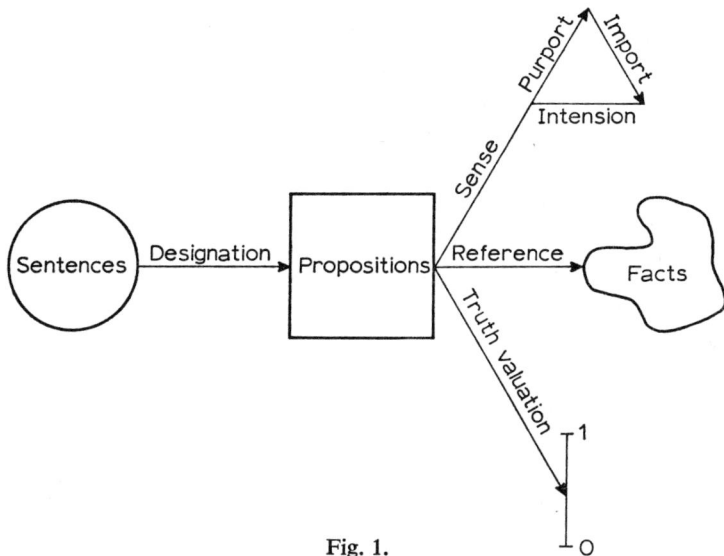

Fig. 1.

respected by the latter. Only scientific theories make 'ontological commitments', or rather assumptions. And they need an ontologically neutral logic and mathematics.

Figure 1, restricted to statements, displays the architecture of our semantic theory.

IV. AN APPLICATION

Imagine a theory of drives or urges, such as hunger, that assumes the intensity D of every drive to be a certain function of some physiological misalignment or imbalance i. More precisely, assume D to be nil below a certain threshold i_0 and to have a sigmoid shape above i_0. One of the infinitely many functions that will comply with this loose description is the function D from reals into reals such that

$$D(i) = au(i)\, i^2/(b + i^2) \qquad\qquad H$$

where a and b are positive real numbers, and $u(i)=1$ for $i>i_0$ and 0 for $i \leqslant i_0$. The above is just the central hypothesis of our bogus theory. The remaining fundamental statements in the theory spell out exactly what the various symbols designate, what the corresponding constructs are about, what if anything they represent, and what their dimensions and units are. For example, there will be a statement to the effect that the domain of the independent variable i is a certain class M of organisms, say humans, while the range of i is the positive real line. On the other hand the theory will contain no indication concerning its own test. In particular it will not contain hypotheses serving to objectify and measure the drive intensities $D(i)$. In principle there are several such objectifiers or indices, either physiological or behavioral, hence several possible techniques of measurement. Usually it is up to the ingenuity of the experimenter to conjecture, test, and use any such relations between covert qualities and their manifestations. In any event, the objectifiers and their measurement are relevant to the test for the truth of the theory, not to its meaning.

A cursory semantic analysis of the quasitheory sketched in the preceding lines yields the following results.

Reference class of i = Reference class of $D = M$ (mankind)

Factual interpretation of i = i represents a physiological imbalance of a certain type (e.g., deficiency of sugar in blood).

Factual interpretation of $D = D(i)$ represents a drive or urge of a certain type (e.g., hunger) as felt by an organism of the kind M suffering imbalance i.

Sense of i = The set of physiological formulas in which i occurs.

Sense of D = The set of formulas entailing or entailed by H and its companions.

Gist of i = The basic (postulated) formulas among all those containing i.

Gist of $D = \{H,$ The above factual interpretation of $D\}$.

So much for the *postulated* sense and reference. Now for the *derived* meanings. They are inferred from the preceding plus an analysis of the mathematical roles the constructs concerned play in the central hypothesis H.

Reference class of a = *Reference class* of $b = M$.

Factual interpretation of a = Maximum drive strength.

Factual interpretation of $(1/b)$ = Strength of the curbing (inhibition) of further drive increases.

Once meanings have been assigned we may proceed to find out truth values on the basis of some body of empirical evidence E relevant to the central hypothesis H. Thus we may pronounce H almost true if $V(H \mid E) = v$ comes close to unity. Whether a statement such as the preceding one is to be called a *truth condition*, is a matter of taste. In any case it is a far cry from a truth condition in elementary logic. And it contributes nothing to the meaning of the theory.

V. CONCLUDING REMARKS

Our program is ambitious, as is any attempt to match life (in our case real science) with virtue (e.g., exactness). We want our semantics to be not only *simia mathematicae* but also *ancilla scientiae*: built *more geometrico* and at the same time relevant, nay useful, to live science. The goal of exactness may sound arrogant but is actually modest, for the more we rigorize the more we are forced to leave out of consideration, at least for the time being. As to the service intention: we should try to be of some help to science because the latter faces semantic problems but has no tools of its own for solving them. If it had such tools scientists would not engage in spirited polemics over matters of sense and reference, as they often do. Witness the debates on whether the relativistic and quantum theories are concerned with sentient observers, whether population genetics refers to populations taken as wholes, whether psychology is actually concerned with the brain, and whether the sense of a theory is excreted by its mathematical formalism or is determined by the way the theory is tested.

A semantics of science should help settle these and similar issues. Moreover it should give sound advice as to how to formulate scientific theories so as to avoid such imprecisions and ambiguities as may give rise to debates of the kind. Constructing such a semantics, both exact and relevant to science, should be more rewarding than either manufacturing neat but irrelevant theories or pursuing erratic polemics on meaning and meaning changes.*

McGill University, Montreal

NOTE

* Research supported by a Killam grant awarded by the Canada Council.

PART III

EROTETICS

NUEL D. BELNAP, JR.

S-P INTERROGATIVES*

I

By an '*S-P* interrogative' I mean an English interrogative something like one of the following four:

(1)   Which S is a P?
(2)   Which S's are P's?
(3)   What's (an example of) an S which is a P?
(4)   What are some (examples of) S's which are P's?

S-P is supposed to suggest 'subject-predicate', but since I use 'subject' and 'predicate' in other ways, it will reduce confusion if I stick to just '*S-P*'. Let us call S the S-term and P the P-term of (1)–(4). I want to marshall three quite different apparatuses for your consideration as formal explications of *S-P* interrogatives, paying particular attention to the way in which S-terms and P-terms enter differently into the logic of the situation. Throughout, the primary aim will be to give an account of these questions in terms of their answers – 'answers' in the sense of 'possible answers' or 'what counts as an answer' rather than in the sense of 'true answers'.

The first interrogative form I want to introduce is a species of *absolute interrogatives*. The key feature of absolute interrogatives is that what counts as an answer thereto is defined on sheerly syntactic grounds; it is a matter of grammar and nothing but grammar. (Later we shall find out that what counts as an answer to what we shall call 'relativized interrogatives' depends on semantic considerations.)

The particular species is that of *which-interrogatives*. (My view of these interrogatives is discussed in detail in Belnap (1963). But I have changed my mind, my terminology, and my notation in the course of preparing Belnap (1968).) A which-interrogative on my account consists of two parts, called the *subject* and the *request*. Let me note in passing that this terminology, though defensible, is made up out of whole cloth. I shall use it to explicate the similarities and differences among (1)–(4), suggesting that these interrogatives are alike in their subjects, but differ in their request.

Using the standard move from English common nouns (e.g., 'man') into formal open sentences (e.g., 'x is a man'), the common subject of (1)–(4) is to be represented by

(5) $(Sx//Px)$,

the surrogate of the S-term appearing on the left of the double virgule, and that of the P-term on the right. What such a subject does is to present a range of alternatives from among which the respondent is to select the material with which to construct an answer. We shall call Px the *matrix* of (5). Its job is to be the matrix from which the alternatives presented by (1)–(4) are derived by substitution of singular terms for the x in Px. We shall say that Sx is the *category condition* of (5). Its function is to determine which singular terms are eligible for substitution in Px in order to obtain an alternative. Unlike Px, Sx sometimes does not appear in the answer at all. (5) may be called a *categoreally qualified* (see Åqvist, 1965) subject. Consider, for example,

(6)   Which positive integer is a prime between 10 and 20?

with formal subject

(7)   (x is a positive integer //x is a prime between 10 and 20).

Here the alternatives – what the interrogative is in some sense 'about' – are '1 is a prime between 10 and 20', '2 is a prime between 10 and 20', etc., where these alternatives are to be described as arising from the interaction of the category condition 'x is a positive integer' with the matrix 'x is a prime between 10 and 20' in the following way. As I stated above, the matrix 'x is a prime between 10 and 20' provides a matrix from which the alternatives are to be generated by substitution of terms for x, while the category condition 'x is a positive integer' controls which terms can be substituted for x – in this case, just the terms '1', '2', etc. Substitution-instances like 'π is a prime between 10 and 20' are not admitted precisely because 'π' is not in the appropriate category suggested by 'positive integer'. The variable x occurring in Sx is bound in (5). Since it is the queried variable, I call it the *queriable* of (5).

If (1)–(4) are alike in their subjects and in the alternatives they present, how are they different? Clearly they *do* put different questions, for what would count as an answer to one would not to another. I arbitrarily call the respect in which they differ their 'requests'. Then I would say that their requests have two dimensions so that the four arise by cross-classification: (1) and (3) are alike in requesting that a single alterative be selected in any possible answer, while (2) and (4) in contrast put no limitation at all on the number of alternatives to be selected; and (1) and (2) are alike in requesting that in each answer a claim be made – I call it a 'completeness-claim' – that the list of alternatives selected in that answer are complete, while (3) and (4) make no such request. I have elsewhere proposed the following symbolism and nomenclature for these four kinds of interrogatives. (Parentheses around the request (left half) as in Table I, are hereafter omitted. The subscript and superscript respectively stand for the

TABLE I

	Requests completeness-claim	Does not request completeness-claim
Requests selection of single alternative	*Unique-alternative* $?(_1^1\forall)\ (Sx//Px)$	*Single-example* $?(_1^1-)\ (Sx//Px)$
Does not request limitation on number of alternatives selected	*Complete-list* $?(_1^-\forall)\ (Sx//Px)$	*Some-examples* $?(_1^- -)\ (Sx//Px)$

lower and upper limits on the number of alternatives allowed to occur in any answer, with '−' signifying the absence of upper limit. Whether or not a completeness-claim is called for is signalled by the presence or absence of '∀', its absence being marked by a '−'. In the full apparatus, other subscript-superscript pairs are allowed, and room is made for a 'distinctness-claim' to the effect that the various substituted terms stand for distinct entities, but these complications are not relevant here since I am chiefly interested in the subject rather than the request.)

The definition of the crucial relation 'A is a direct answer to I' is given for these four sorts of interrogatives by the following schemata:

Single-example: Pa
Unique-alternative: $Pa \,\&\, (x)(Sx \supset (Px \supset x=a))$
Some-examples: $Pa_1 \,\&\, \ldots \,\&\, Pa_n$
Complete-list: $[Pa_1 \,\&\, \ldots \,\&\, Pa_n] \,\&\,$
     $(x)(Sx \supset (Px \supset x=a_1 \vee \cdots \vee x=a_n))$.

The universally quantified conjuncts of the unique-alternative and complete-list answer have the effect of claiming that the selection of alternatives given is complete: the whole truth. Although the category condition Sx does not appear in every answer, still its presence is felt throughout by way of the following requirement:

(R)   The names $a, a_1, \ldots, a_n$ must be in the category – since they are names we shall say *nominal* category – determined by Sx.

It is part of the explication that whether or not a name does in fact belong to Sx's nominal category shall be entirely a grammatical matter and effectively decidable. It is also essential to a correct understanding of this explication that Sx be viewed not as an arbitrary predicate but rather as one from among a list of predicates – I call them *category-conditions* – antecedently given by the grammar of the language. Appropriate candidates in English (or middle English) would be 'x is an integer', 'x is a man', and 'x is a country', and cases similar to these in respect of allowing more or less straightforward and more or less non-empirical determination of which instances are true. Of course in our *formal* reconstruction we can strike out the 'more or less' clause. (The use of category-conditions is intended as a flexible and easily applicable variant on the use of a many-sorted logic.)

II

The second way of handling S-P interrogatives such as (1)–(4) is based on a completely different underlying idea. Instead of presuming that these interrogatives have exactly the same answers (in the sense of what-counts-as-an-answer) regardless of the state of the world, one supposes that answerhood is semantically relativized to world-states. (The notion of relativized interrogatives was inspired by some features of Åqvist (1965).

The present account overlaps that in Belnap (1969). In order to avoid pointless rewording, a few paragraphs have been lifted from that paper.) Since the idea is independently interesting, we develop it for a while without reference to (1)–(4). We shall need to invoke the single-example whether-interrogative notation $?_1^1-(A_1,...,A_n)$, which is to be taken as an absolute interrogative the answers to which are just $A_1,...,A_n$.

The need and point of relativized interrogatives, as I call them, is most clearly seen in the case of *conditional* interrogatives such as

(8)   If you are going, are you taking your umbrella?

This is to be distinguished from a *hypothetical interrogative* like

(9)   If you were to go, would you take your umbrella?

since (8), unlike (9), does not call for answers having the form of a conditional (e.g., 'If I go, I'll take my umbrella'), but rather asks that an answer, itself unconditional in form, be supplied only if a certain condition is true. The intent of (8), as we read it, is that an answer is called for only if the respondent is going; if he is not going, then though he may take it upon himself to so inform the questioner, to do so is not directly called for.

One difficulty in generalizing from such examples is that English has no short and idiomatic interrogative way of making the distinction we require, since, as it seems to me, (8) *could* be taken as a plain hypothetical like (9), or even as an absolute interrogative $?_1^1-(\sim G, U, \sim U)$, with the negation of the condition as an additional direct answer. More unambiguous would be what we take to be an English equivalent of (8): "If you are going, tell me whether you are taking your umbrella." Nevertheless, it seems best to see what happens if we take (8) as a distinctive interrogative form construed along the lines suggested above. Evidently, then, (8) cannot be explicated as an absolute interrogative. We therefore require the following crucial pair of locutions, in which answerhood is relativized to world-states. (World-states can be identified with models or interpretations à la Tarski, or with the possible worlds of contemporary modal logic. Since the present development does not utilize modal operators, I temporarily make the former identification with interpretations.)

(i)   *I* calls for an answer in world-state *M*.
(ii)   *A* is a direct answer to *I* in world-state *M*.

It is understood that the second locution is only defined provided that I calls for an answer in world-state M, I shall use '*is operative*' as a synonym for 'calls for an answer', and say that an interrogative is *inoperative* if it does not call for an answer.

Absolute interrogatives are those for which only the straightforward '*A* is a direct answer to *I*' is defined, while *relativized interrogatives* are those for which direct answerhood is relativized to world states. The distinction between absolute and relativized interrogatives is then seen to depend on which metalinguistic locutions are defined, or in natural language, on which locutions are appropriate. Among the relativized interrogatives, however, we can single out those as *categorical* which have the same direct answers in every world-state and are always operative.

In order to subsume absolute interrogatives under the new concepts, we will say that, where I is an absolute interrogative, the new locutions apply to it in the following ways: (i) I is operative in *every M*, and (ii) its direct answers in every M are the same, namely, those previously defined as its direct answers in the absolute sense. Subsumed absolute interrogatives therefore turn out to be categorical.

Conditional interrogatives are clearly *not* all categorical since sometimes they do not call for an answer. Let us agree to use

(10) (A/I)

as the *conditional interrogative* with *condition A* and *conditioned interrogative I*. For example, the going out/umbrella interrogative (8) would become

(11) $(G/?_1^1 - (U, \sim U))$.

But what does the notation (10) mean? It follows from the above that to say what this interrogative means we must say (i) in which world states it calls for an answer, and (ii) for those in which it so calls, what its direct answers are. Application of these ideas to (10), in accordance with our informal account of (8) as calling for an answer only when G is true and then calling for an answer to the umbrella yes-no interrogative, is straightforward.

(i) (A/I) *calls for an answer in M* iff A
is true in M and I calls for an answer in M.

(ii) Provided (A/I) calls for an answer in M, B *is a direct answer* to (A/I) *in* M iff B is a direct answer to I in M.

For example, 11 calls for an answer if and only if G is true, and provided G is true, its answers are those to $?_1^1-(U, \sim U)$, i.e., just U and $\sim U$. When G is false, the concept of direct answerhood is undefined, and all we can say is that the interrogative does not call for an answer.

An important reason for putting conditions on interrogatives is, to employ a happy expression of Åqvist, to *guard* them: not knowing whether an absolute interrogative has a true direct answer, we may ask, "If it does, please supply one" by putting the claim that it has a true answer as a condition on itself. For example, we may guard "Was it suicide or murder?" as expressed by, say $?_1^1-(S, M)$, by asking "If it was either suicide or murder, which one was it?":

$$(S \vee M/?_1^1 - (S, M)).$$

Suppose that in fact it was neither suicide nor murder but an accident. Then to use the absolute interrogative, "Was it suicide or murder?" would be to do something 'bad', to call for a true answer when it is not possible to give such. But to use its conditionalization would be acceptable, since in the given circumstances the conditionalization would simply be inoperative, not calling for an answer at all. A particularly entertaining form of this maneuver occurs when the guarded interrogative is a *Hobson's Choice*, i.e., an interrogative with but one direct answer, as $?_1^1-(A)$: 'Tell me that A'. Let us grant that this interrogative form is of no (?) utility. But consider its conditionally guarded cousin,

$$(A/?_1^1 - (A))$$

which calls for an answer if and only if A is true, and then asks for A. In short, it answers to the form, "If it is true that A, tell me so," the nicety of which can be seen from examples like "If you can't hear me in the back row, tell me so."

Let us turn to another way of constructing new interrogatives out of old. Given a set of absolute interrogatives, we can define their conjunction as that interrogative which has as its answers a conjunction containing, for each interrogative in the set, a conjunct which is an answer to that interrogative. Thus an answer to a conjunction of a set of aboslute interrogatives provides an answer to every interrogative in the set. The natural adaptation of this idea to relativized interrogatives is to say that a conjunction of a set of relativized interrogatives has as its answers conjunctions containing, for each *operative* interrogative in the set, a conjunct which is an answer to that interrogative. Thus, an answer to a conjunction of a set of relativized interrogatives provides an answer to every member of the set which calls for an answer. The idea doesn't make much sense for infinite sets of *absolute* interrogatives, conjunctions being only finitely long, but it does make sense for infinite sets of *relativized* interrogatives for the following reason: although the entire set may be infinite, the set of its operative members, i.e., those calling for an answer, may well be finite, so that it is easy enough to concoct a (finite) conjunction with a conjunct answering each operative interrogative in the set. With this in mind, let us skip finite conjunctions of interrogatives and proceed directly to infinite ones via a new interrogative form, called a *universalized interrogative*,

(12) $\forall x I x,$

where we assume x occurs free in the interrogative I, hence not as a queriable. The interpretation is as follows:

(i) 12 calls for an answer in M iff some substitution-instance Ia of Ix does so.

(ii) Provided 12 calls for an answer in M, a formula is a direct answer to 12 in M iff it is a conjunction containing for each operative substitution-instance Ia of Ix, a conjunct which is an answer in M to Ia, and containing no other conjuncts.

For example, let Sx mean that x lies between 10 and 20, and let Px mean that x is a prime. Then

$$\forall x (Sx/?_1^1 - (Px, \sim Px))$$

amounts to the universalized conditional interrogative, "For each x, if x is a number between 10 and 20, is x a prime?" It should cause the respondent to reply with a nine-term conjunction, each conjunct of which directly answers an appropriate substitution-instance $?_1^1-(Pa, \sim Pa)$ of $?_1^1-(Px, \sim Px)$. Here 'appropriate' means that the matching substitution-instance Sa of Sx is true. And just because Sx is true of only nine numbers, the interrogative makes sense. This particular interrogative has 2^9 distinct answers (ignoring order), only one of which, of course, is true. But it is to be noted that expressions containing more or fewer than nine conjuncts, or containing the wrong sort of conjuncts are not just false answers, but non-answers. To insert a conjunct $P(7)$ is not to directly respond to this interrogative, at least not in our world.

We are now ready to apply this machinery to S-P interrogatives, in particular to (2). The idea is that we do not, as before, interpret the S-term as a category grammatically delimiting the set of what-counts-as-an-answer. Rather, its action is semantical via its role in a universalized conditional interrogative. To see better what is happening, let us introduce $*A$ as an abbreviation for the conditionally guarded Hobson's Choice interrogative $(A/?_1^1-(A))$, so that $*A$ is read, 'If A is true, tell me so'. Then 2 – 'Which S's are P's – is on this scheme to be taken as

(13) $\forall x(Sx/*Px)$,

which can be back-translated as 'For each S, if it is a P then tell me so', or perhaps just as 'Which S's are P's?' Formally, the interrogative 13 has the following properties: (i) 13 calls for an answer just in case at least one S is a P; (ii) if $a_1, ..., a_n$ is a complete list of all the S's that are P's, then (ignoring subtleties of order, etc.) Pa_1 &...& Pa_n is 13's one and only answer, an answer which is bound to be true; (iii) though *calling* for an answer even when infinitely many S's are P's, 13 does not in such circumstances *have* any answers. (Such an interrogative I call 'dumb', for that is how a respondent must remain.)

In order to handle single-example interrogatives like (3), one naturally uses a mode of interrogative combination more akin to existential than to universal generalization. By saying that an interrogative is the *union* of a set of interrogatives, I mean in the absolute case that an expression counts as an answer to it just in case the expression counts as an answer to at least one interrogative in the set, and in the relativized case, just in case it counts as an answer to at least one *operative* interrogative in the set. There is no problem here about infinite sets since we are not combining answers but rather taking them one at a time. As a symbol for the variable-binding operation I will use the set-theoretical union sign rather than an existential quantifier, since in contrast to universalized interrogatives, what is going on is a set-theoretical operation on sets of answers rather than a logical operation on the answers themselves.

(14) $\cup xIx$,

called a *unionized interrogative*, is then defined as follows: (i) (14) calls for an answer in M just in case some instance Ia of Ix calls for an answer in M; (ii) an expression is a direct answer in M to (14) just in case it is a direct answer in M to some substitution-instance Ia of Ix.

Then our candidate explication for (3), 'What's an example of an S which is a P?', is

(15) $\cup x(Sx/?_1^1 - (Px))$.

We might back-translate this interrogative by means of 'For some x if it is an S then tell me that it is a P', or perhaps just by 'What's an example of an S which is a P?' taken in the sense, 'Among S's, what's an example of a P?' The formal properties of our candidate 15 are as follows: (i) it calls for an answer just in case there is at least one S; (ii) when operative, (15)'s answers each have the form Pa, where in fact a is an S. Of course if one wants to be sure of calling only for true answers, one could use instead a conditional Hobson's choice, $\cup x(Sx/*Px)$.

One interesting feature of unionized interrogatives is that in spite of the fact that $\cup x$ is very much like an existential quantifier, and in spite of the fact that we have been trained to avoid putting existential quantifiers in front of conditionals, it turns out that $\cup x$ sits very well in front of a conditional (Sx/Ix), giving exactly the desired effect of saying 'For some x such that Sx, answer me Ix'.

Unionized interrogatives have uses other than explicating (3). For example,

$\cup x(Sx/?_1^1 - (Px, \sim Px))$

could be used to ask, 'For some x between 10 and 20, is x a prime?', or more colloquially, 'Tell me for some number between 10 and 20 whether or not it is a prime'. And a version of (1) could also be given by a unionized interrogative:

$\cup x(Sx/(Px/?_1^1 - (Px \ \& \ (y)(Sy \supset (Py \supset y = x)))))$

would represent the version in which it is not presupposed that there is at least one S which is a P, but it is presupposed that there is at most one S which is a P. In this version the uniqueness claim occurs as part of the answer. In contrast,

$$\cup x (Sx/[(Px \;\&\; (y) (Sy \supset (Py \supset y = x))]/?_1^1 - (Px)))$$

would be a version in which neither existence nor uniqueness is presupposed, and in which the uniqueness is used as a condition on answerhood but does not itself appear as part of the answer. One could also substitute $\forall x$ for $\cup x$; this is always possible when, as in this case, at most one of the ingredient interrogatives can be operative. A third version,

$$\cup x (Sx/?_1^1 - (Px \;\&\; (y) (Sy \supset (Py \supset y = x)))),$$

treats 1 exactly as a special case of (3), asking for an example among the S's of a P unique among the S's.

It does not seem possible to treat (4) with the devices at hand, nor does it seem possible to give alternative versions of (2) analogous to the above versions of (3). In both cases the trouble is that the answers contain elements which are not schematizable by first order matrices.

One way among several that these problems could be solved would be by introducing a variable, F, ranging over finite sets $\{a_1, ..., a_n\}$, with the understanding that $y \in \{a_1, ..., a_n\}$ iff $y = a_1 \vee ... \vee y = a_n$. Then the some-examples interrogative (4) could be given by

$$\cup F \forall x (x \in F/(Sx?_1^1 - (Px))$$

which will then have as answers every conjunction

$$Pa_1, Pa_1 \;\&\; Pa_2, ..., Pa_1 \;\&\; ... \;\&\; Pa_n, ...,$$

such that $a_1 ..., a_n$ are all S's. And a version of (2) in which the completeness-claim is made part of the answer and not just a condition on answerhood would be given by

$$\cup F ((x) (x \in F \supset Sx)/^*((x) (x \in F \equiv Px))),$$

which would have as answers the formulas

$$(x) (x \in \{a_1, ..., a_n\} \equiv Px\}$$

such that $a_1, ..., a_n$ is a complete list of all P's among the S's. This way is simple, and uses sets only in ways of which no nominalist need be ashamed.

III

Another way to solve the problems of explicating (4) and of giving alternate versions of (2) would be to introduce relativization on a different basis. In all the above examples the ground floor of relativization was occupied always by the conditional interrogative. Instead one can introduce relativization into the very subject ($Sx//Px$) of a which-interrogative itself. Since the function of a subject is to present a range of alternatives from among which the respondent is to make a selection, a relativized subject will in general present different alternatives in different world-states. Thus, for each relativized subject, σ, we must define the following two fundamental locutions:

(i) σ *is operative* in interpretation M (which is to say that what alternatives it presents is defined); and provided σ is operative in M,

(ii) *A is an alternative presented by σ in M.*

The ground floor relativized subject will be

($Sx///Px$),

with the understanding that (i) it is always operative and (ii) it presents as alternatives exactly those instances Pa_i of Px such that Sa_i is true. (Another option would be to make ($Sx///Px$) operative only when there are some S's, or even only when there are S's which are P's.) Such a subject is said to be a *conditionally qualified* subject, in contrast with the categoreally qualified subject ($Sx//Px$).

For the four interrogative forms just like those in Table I, except for containing conditionally qualified instead of categoreally qualified subjects, relativized answerhood is defined in *exactly* the same terms with just one exception: the requirement (R) of I, which said that the names used in substituting into the matrix Px must be in the nominal category determined by Sx, is changed to

(R') The names $a, a_1, ..., a_n$ must be such that $Sa, Sa_1, ..., Sa_n$ are true in M.

It is understood that for conditionally qualified subjects ($Sx///Px$) there is no restriction on the choice of Sx, as there is in the case of categoreally qualified subjects ($Sx//Px$). Nor is there any presumption that one can effectively decide whether or not Sa is true. It follows that, as for relativized interrogatives in general, what counts as a direct answer is no mere matter of grammar, but instead depends on the facts.

IV

Having developed these differing explications of the *S-P* interrogatives (1)–(4), let us draw some comparisons. We begin with some remarks on the which-interrogatives introduced in I in explication of (1)–(4).

(1) They are *absolute interrogatives* in the sense that answerhood is not relativized to the state of the world but is instead a purely grammatical matter.

(2) Because answerhood is grammatical, it can be *effectively decidable*; both questioner and respondent using this apparatus can effectively tell which pieces of notation are and which are not direct answers so that the questioner never has to ask, "Was that an answer to my question?"

(3) They are *subject-request interrogatives* in the sense that answerhood is defined by means of the interaction of a subject, which presents alternatives, and a request which lays down specifications as to how these alternatives are to be combined in order to form an answer.

(4) Their subjects are *categoreally qualified*, which means that the *P*-term of (1)–(4) is used as a matrix in which to substitute and that the *S*-term of (1)–(4) shows up as a category condition used to limit in an effective way what is to count as an alternative and therefore, derivatively, what is to count as an answer. There are two reasons for designing a formal apparatus with this feature.

(a) In the first place, the formal apparatus is to be used in an explicatory way, and since one finds the feature in English, there is a *prima facie* case for including it in the formalism. The point from English is that given 'What's an example of a positive integer lying between 10 and 20?' one counts something like 'General Sherman is a prime lying between 10 and 20' not as a false answer, but rather as recognizably not an answer at all, easily distinguishable from say '8 is a prime lying between 10 and 20', which is just false. The General Sherman response is, relative to the interrogative, a category mistake, though it might not be a category mistake relative to an interrogative with a wider category condition. (I believe category mistakes *generally* to be relative to interrogatives; there is a good story here.)

(b) Second, the formalism is to a certain extent to be taken normatively, so that its features are to be judged as useful or not independently of their occurrence in English. And being able to describe with maximum and effective specificity the area in which answers are to be found is of benefit to all parties to an interrogative transaction. It is desirable from the point of view of the respondent since the narrower the category condition which is used, the narrower will be the field in which he has to search for an answer – if search he must. And it is desirable from the point of view of the questioner since he can use the machinery to limit what-counts-as-an-answer to those responses he would find helpful. Thus, one *could* use the category apparatus to exclude 'The author of Waverly is the author of Waverly' as an answer to 'Who is the author of Waverly?', using instead a category predicate including in its nominal category only (say) recognizably proper names.

Passing now to the interrogatives introduced in II, we may make the following remarks.

(1) These interrogatives are not absolute but *relativized* in the sense that what counts as a direct answer thereto is relative to the state of the world.

(2) Because answerhood is semantical, it cannot be expected to be effectively decideable. For example, without knowing the truth-value of the condition A of a conditional interrogative (A/I), one cannot tell whether an answer is called for or not. This feature seems to me accurately to mirror certain aspects of the situation in English. For example, given 'Robinson Crusoe is a conductor of the Vienna Philharmonic Orchestra', we may well feel that it is an empirical matter whether or not one should count it as an answer to 'Which Alpine guides are conductors of the Vienna Philharmonic Orchestra?' – just as empirical as the matter of its truth.

(3) These interrogative forms are not subject-request. Rather they grow out of combining the 'tell-me-that-A' form, the conditional interrogative, and either universalization or unionization. To put it as a mouthful, they are generalized conditional Hobson's Choices. These explications gain interest from the fact that the ingredient forms are both simple and independently interesting, and that the mode of combination sounds intuitively right.

(4) As in the case of the interrogatives of I, the *P*-term is, as above, used as a matrix in which to substitute, and also as above, the *S*-term is used to limit what is to count as an answer. But the limitation is semantical rather than grammatical, and by so much not effective.

I have suggested that I could see both an explicative and a normative point in introducing the categoreally qualified interrogative. I have also allowed the explicative force of relativized interrogatives as answering to deep-seated features of our natural language. But I cannot at this particular stage of investigation claim any normative value for the machinery of relativization. For example, if it is to be an empirical matter whether the sentence 'Robinson Crusoe is a conductor of the Vienna Philharmonic Orchestra' does or does not count as an answer to 'Which Alpine guides are conductors of the Vienna Philharmonic Orchestra?', then I cannot (right now) see any point in differentiating between calling that sentence a non-answer and a false answer. But I remain diffident about this negative judgment, and in any case I am reasonably certain that the picture changes when temporal parameters are added to the machinery governing *when* an answer is wanted. There is, I think, no substitute for the tensed conditional reading of 'If there is a fire, then (i.e., at that time) tell me where the nearest exit is located'.

(5) Combining generalization, conditionalization, and whether-interrogatives, including Hobson's Choices, does not seem to give an adequate account of some-examples interrogatives, nor of readings of complete-list interrogatives which require the completeness-claim to be part of the answer.

Let me now say just a few words about the interrogatives introduced with considerable brevity in III.

(1) They are relativized interrogatives, thus in this respect like those of II and unlike those of I.

(2) They are subject-request interrogatives, in this respect being like those of I rather than those of II. Because of this they can be used to explicate in a relativized way the some-examples interrogative (4) as well as those versions of the complete-list interrogative (2) resistant to explication as generalized conditional Hobson's Choices.

(3) Their subjects $(Sx///Px)$ are conditionally qualified rather than categoreally qualified like $(Sx//Px)$ of I. This means two things when this form is regarded as the target of the *S*-term and *P*-term of some English *S-P* interrogative.

(a) Unlike the categoreally qualified subject ($Sx//Px$), there is no grammatical restriction on the choice of Sx. In particular, Sx need not answer to some common noun which we would naturally regard as a category-expression in English and for which we could lay down more or less effective conditions for truth of substitution-instances. For this reason this form is doubtless a better explication of an S-P interrogative like 'Which Alpine guides are conductors of the Vienna Philharmonic Orchestra?', for the S-term of this interrogative is far from categoreal. But the other form is probably better for 'Which men are conductors of the Vienna Philharmonic Orchestra?', in which the S-term is indeed categoreal.

(b) In exchange for removing the restrictions on the choice of Sx, one has to pay the price of non-effectivity. The nature and significance of this price are to me at this moment obscure, though I have no doubt further research will provide further light.

Univ. of Pittsburgh, Pittsburgh

NOTE

* This research has been supported in part by the System Development Corporation, Santa Monica, California, and in part by the National Science Foundation, Grant No. GS-28478. I wish to thank Thomas Steel, with whom many of the key ideas were worked out. An early version was read in 1968 at the Oberlin Conference. I have profited from comments there given by L. Åqvist.

BIBLIOGRAPHY

Åqvist, L., 1965, *A New Approach to the Logical Theory of Interrogatives, Part I: Analysis*, Uppsala.

Belnap, N. D., Jr., 1963, 'An Analysis of Questions, Preliminary Report', Santa Monica, System Development Corporation, California.

Belnap, N. D., Jr., 1968, *Erotetic Logic*, mimeographed. (A portion of a study to be authored jointly with Thomas Steel.)

Belnap, N. D., Jr., 1969, 'Åqvist's Corrections-Accumulating Question-Sequences', in J. W. Davis, D. J. Hockney, and W. K. Wilson (eds.), *Philosophical Logic*, D. Reidel, Dordrecht, Holland.

PART IV

PHILOSOPHY OF MATHEMATICS

WILLIAM S. HATCHER

FOUNDATIONS AS A BRANCH OF MATHEMATICS

I. PROLOGUE ON MATHEMATICAL LOGIC

It is quite clear the way in which we can, and perhaps should, consider mathematical logic as a branch of mathematics, in fact a part of applied mathematics. The main technique used in applied mathematics is that of studying a phenomenon by building a mathematical model of it. Such a model is an idealization which approximates the phenomenon by concentrating on a few relevant features and ignoring what seems to be less significant. If the model turns out to have high predictive and explanatory value, at least in some useful situations, it is subjected to extensive theoretical development and becomes a 'theory' (i.e. probability theory, heat theory, etc.).

It is precisely in this way that mathematical logic can be viewed as applied mathematics. The phenomenon which underlies mathematical logic is common-sense logical inference. This phenomenon has several aspects, in particular, language, 'reality', meaning, and deduction. The mathematical model of this situation consists of the following abstractions: formal languages for language, mathematical structures for reality, interpretation in structures for meaning, and formal deduction for deduction. Though this model neglects many features of the actual phenomenon (most notably the psychology of the reasoner), it has been judged useful enough to be studied extensively.

Now, if one accepts the above as a reasonable summary of what mathematical logic is, then mathematical logic is no more 'philosophical' than any other branch of applied mathematics. The only reason one might be tempted to argue otherwise is because in logic it is, in part, the human thought process itself which is being modelled, thus apparently rendering the model more 'subjective'. Yet, if modern philosophical analysis has shown anything, it has shown just how much subjectivity enters into our models even of physical phenomena. Thus, logic would not seem to have any special status on this account. One could even argue that logic is more objective than physics since it is more explicit about the point where subjectivity enters in.

Moreover, if one takes a thoroughgoing pragmatic viewpoint, there is not even a necessity to try to distinguish absolutely between that part of our world view which comes from the viewer and that part which derives from the thing viewed. We have only to evaluate, through experience and usage, whether or not a given model of a given phenomenon has explanatory value.

Of course, there are philosophical problems concerning the nature of our knowledge and the ultimate justification for using any given mathematical model. But these problems obtain equally for all phenomena and their models and not peculiarly for logical phenomena. Thus, it is not logic which is philosophical but rather the whole question of how we know.

II. MATHEMATICS AND KNOWLEDGE

It is certainly true that epistemological questions have been influenced by mathematics. The tradition of this influence goes back in a strong form at least as far as Plato, and everyone is aware of the central role played by mathematics in the analytic philosophy of the twentieth century. But if mathematics has influenced epistemology, it is also fundamentally clear that mathematics, because it claims to know, must itself depend on the philosophically prior question of how we know anything at all. It makes no sense to elaborate an epistemology of mathematics apart from the elaboration of epistemology *tout court*. The basic problem, then, is how to situate mathematics with respect to knowledge in general.

Of course, we should beware of supposing *a priori* that 'mathematics' designates a clearly delineated part of our knowledge and that we have only to find the right objective criteria for characterizing it. This attitude begs the question and departs from a pragmatic approach. Let us pose the problem rather in these terms: "Is it useful, and to what degree, to classify a part of our knowledge as 'mathematical' and what are 'good' criteria for doing so?"

I feel that the most fruitful way to regard mathematics is to consider it to be the exact part of our thinking, our thinking about anything.[1] Whenever we objectify, abstract, and make precise our thinking we are doing mathematics on some level. From this view, it follows that there is no specifically 'mathematical' intuition. Anything is potentially a source of our intuition.

This view is certainly supported by modern developments in mathematics in which non-numerical mathematical structures are widely applied to previously 'unmathematized' areas; and no one seriously doubts that there will be new and important applications in the future.

The process of precising, abstracting and objectifying involves the use of various conceptual, linguistic, and logical tools. We should, however, beware of trying to explain the relationship between these tools in any too-simple way. In all of our thought, there is a give and take between the conceptual and the formal, the intuitive and the logical. A concomitant of the view of mathematics that I am urging is that this duality is not a divorce and that neither aspect is prior to the other or dominates the other. I feel that it is wrong to consider that mathematics has any clearly-defined starting point or that mathematics is basically founded either on intuition or on a formalism of some sort. Rather, both of these tools combine to create mathematics. A similar view is expressed in a recent paper of Lawvere (cf. Lawvere, 1969, p. 281).

The classical modern approaches to foundations, however much they may differ in their conceptions or their formulations, all share a common desire of giving a 'once and for all' foundation for mathematics. Logicism sought to found mathematics on our intuition of purely logical relationships. Intuitionism sought to found mathematics on our intuition of the concrete (one is tempted to say our abstract intuition of the concrete). Formalism sought to found mathematics on a certain kind of formal system.

Each one of these approaches has subsequently been seen to have important drawbacks due to various kinds of inadequacies which it exhibits. Some of these inadequacies are heightened by the well-known incompleteness and undecidability results as well as certain logical antinomies. Of the three basic approaches, the one which was most philosophically attractive in its initial formulation was logicism. For if, in fact, it could be shown that our mathematical intuition flows from our logical intuition, considerable clarification of the epistemological process would result. This undoubtedly accounts for the tremendous attraction logicism held for strong thinkers such as Carnap, even when evidence against logicism continued to pile up.

Since logicism is perhaps the strongest thesis in support of a 'once and for all' foundation for mathematics, I want here to reconsider it briefly.

III. LOGICISM REVISITED

The first formulation of logicism was perhaps also the most precise since Frege not only formulated his philosophical ideas clearly but was highly adept at technical manipulation and formal rigor. If we assume predicate logic as given (and part of mathematics), then the two further axioms of Frege's system are extensionality and abstraction. The underlying philosophy of the system could be described as follows: "When we formulate clearly the purely logical relationships between predicates and their extensions, then we have, *de facto*, mathematics." The essence of this intuition was the abstraction principle which says that every predicate has an extension which is an element of the universe.

Now, Russell's paradox, which showed that the abstraction principle was false in general, did not in itself destroy this philosophy. Russell showed only that Frege's formulation was not correct. Philosophically speaking, he showed that Frege's intuition of the logical relationship between predicates and extensions was innacurate. Russell's reformulation using type theory was intended to recapture, correctly this time, the intuition which Frege had badly formulated. Although the axioms of choice and infinity were independent with respect to type theory, one could argue, as Carnap did, that these should simply be taken as explicit hypotheses in theorems depending on them, but that the abstraction principle (in type theory) was still the basic intuition on which to found mathematics (cf. Carnap, 1970, pp. 345–6).

Indeed, in much of the current literature on foundations, one still senses a longing for Frege's system. Witness, for example, A. P. Morse who evaluates his own system as one which captures "... the intuitive simplicity of Frege's beautiful but inconsistent system." (Morse, 1965, p. xxvii). As is well known, Morse's system has a particularly strong form of the abstraction scheme as a basic principle of class generation.

What is it in the abstraction principle which leads authors again and again to feel that it is so basic? Undoubtedly it is, in large measure, the astonishing use which Frege and Dedekind were able to make of this apparently simple principle. They showed, by explicit constructions, how all current mathematical notions were to be defined as extensions of predicates of a purely logical sort. Thus, even though other mathematicians might feel that their spontaneous intuition of, for example, the natural numbers was quite different from the logical one urged by Frege and Dedekind, they could not but admit that the Frege-Dedekind constructions had succeeded in generating the natural numbers on the basis of the abstraction principle.

Still, there was a basic tension between the spontaneous or 'natural' intuition of mathematical objects such as the natural numbers or the real numbers, and the constructions based on one global logical intuition. This tension still prevails today. For example, the natural intuition of a function is that of an operation whereas the usual logical definition constructs functions as sets of ordered pairs. Philosophically, then, the real question posed by logicism is: does our natural intuition of those structures needed for mathematics on the one hand and our logical intuition of the abstraction principle on the other lead us in the same direction? More simply, is this tension real or imagined, essential or marginal?

There is one current notion of mathematics which cannot be defined in any evident way from the abstraction principle, namely that of a choice set. No attempt was ever made by Frege or Dedekind to obtain the choice principle from the abstraction principle.[2] Of course, by the time that Zermelo had explicitly formulated the choice principle and focussed attention on it, Russell's paradox was already known. The independence of the choice principle in type theory left the question unanswered.

However, one could argue that type theory is unnecessarily restrictive because there is not a complete ambiguity of types. This can be seen from the fact that there are sentences S in type theory such that the type lift S^+ of S (obtained from S by increasing the type of each variable one unit) is provable while S is not provable. Complete typical ambiguity can be obtained by adding the axiom scheme $S \equiv S^+$, where S is any sentence of type theory. Such a system approximates even closer our intuition of the abstraction principle.

How does the axiom of choice fare in this system? Specker has answered this question by showing that the axiom of choice is false in type theory with complete typical ambiguity (Specker, 1962). Thus, by approximating more closely our logical intuition of the abstraction principle we have actually contradicted our natural intuition of the choice principle. In other words, the tension between our two sorts of intuition is seen to be essential and not marginal. The two intuitions lead us in different directions.

Let us note in passing that the axiom of infinity is provable in type theory with complete typical ambiguity.

The above example would seem to say that logicism is mistaken even as a philosophical ideal, apart from the other technical objections resulting from the various incompleteness results of Gödel and Tarski. Indeed, the theory of types with complete typical ambiguity is equivalent to Quine's New Foundations (cf. Specker, 1962). This latter system was originally posed as a purely formal generalization of type theory, without any model or interpretation in mind. Though it originally found favor with logicians such as Rosser, it has more recently come to appear as something of an oddity.

It is interesting to contrast all of this with Zermelo's system which was originally an *ad hoc* attempt simply to list, without any philosophical justification, those principles actually used or needed by mathematicians. In other words, Zermelo's approach was more in the line of a formulation of spontaneous mathematical intuition rather than that of deriving notions from one global logical intuition. It may be fairly said that almost every system of set theory seriously studied and used by mathematicians in recent years has been a natural extension or modification of Zermelo's.

IV. FOUNDATIONS AS A BRANCH OF MATHEMATICS

The failure of logicism and the various inadequacies which continue to appear in connection with each of the traditional approaches to foundations strongly suggest that we should renounce the attempt at giving a 'once and for all', absolute foundation to mathematics. It is my view that we should, in fact, accept this suggestion. I feel that the relative approach to foundations allows us to put mathematics into perspective.

Philosophically, we return to a healthy pragmatic openness in which we view mathematics and logic in less of a special and exalted way. Their importance still remains, but their philosophical dependence on general epistemology is frankly recognized and accepted.

Mathematics, the exact part of our thinking, becomes a phenomenon to be studied. We have an intuitive notion of the process of abstracting, precising and objectifying. Moreover, this process uses the deductive techniques and semantic interpretations which are used generally in our thinking. These latter techniques already have a mathematical model in mathematical logic. Since we know already from the various incompleteness theorems of logic that we cannot formally and consistently capture the total intuition of mathematics, we must content ourselves with various languages which capture, in a natural and consistent way, an important part of our intuition. Such a language I call a *foundational system*.

A foundational system is a mathematical object. The comparative study of foundational system is clearly a branch of mathematics, more particularly a branch of mathematical logic. I call this branch of mathematics *foundations*. Such a branch of mathematics does not need more (or less) philosophical justification than any other branch of knowledge or than mathematics itself. But, to repeat, the whole process derives its philosophical justification from our whole epistemology.

The idea of naturalness mentioned above is an important intuitive criterion in elaborating foundational systems. A foundational system should not just 'passively' capture a large portion of mathematics, but it should also lead and stimulate our intuition. It should give us new ideas about how to objectify and abstract.

The recently developed theory of categories illustrates this point very well. Set theory was originally formulated as a tool to clarify certain notions in analysis. Analysis was essentially the study of certain number systems which were themselves powerful abstractions. As set theory was developed in its own right, the structural approach to mathematics become more and more important. But as our intuition of structure grew stronger set theory appeared to be less and less natural since the important feature of a given structure are often irrelevant to its set-theoretical properties which follow from the membership relation. Category theory has shown itself to be extremely useful in clarifying the notion of structure and in symplifying and abstracting the relevant features of structure via universal mapping properties. This clarity shows the naturalness of category theory. To take an example: the category-theoretic definition of the set of natural numbers as a universal object (essentially the free Peano algebra on one generator) is significantly different from the set theoretic intuition of the numbers as counting sets. Lawvere's characterization of set exponentiation as right adjoint to product is another example. Moreover, since both of these definitions are special cases of the category-theoretic notion of universality, we are led to think about these traditional notions in new ways. This leads to new kinds of theorems and the intuiting of new kinds of relationships.

In fact, Lawvere has shown how to develop a foundational system based on category theory. The most recent form of this system, called a Topos, is considered by many to be the most fruitful foundational system to be developed in the last years. Considerable simplification and clarification of logic, set theory, and algebraic geometry have already been achieved.

In sum, foundations can be considered a branch of mathematics in very much the same way that group theory or topology are considered branches of mathematics. A branch of mathematics is characterized by the kinds of structures studied and the kinds of questions treated. In the case of foundations, the structures studied are foundational systems. The kinds of questions treated are, for example, the relative strength of various systems, independence of axioms, consistency, various kinds of adequacy, etc.

Traditionally, 'working' mathematicians such as group theorists or topologists have assumed that a once and for all foundation for mathematics had already been given somewhere by some logician. For them, foundational questions were relegated to some nether land to be conveniently forgotten. It is therefore interesting to observe that foundational questions are now being more frequently raised in all branches of mathematics. This is due in part to the fact that new techniques have forced researchers to consider questions which have different answers within different foundational systems. (For example, many questions in the theory of abelian groups now depend on whether or nor one assumes strong axioms of infinity in set theory.) It is also due to increasing acceptance of the relative approach to foundations whereby foundational systems and their models are studied in the same way as groups and topological spaces have been studied.

Philosophically, the view of foundations I am urging has some of the flavor of Mehlberg's 'pluralistic logicism' (cf. Mehlberg, 1962). Perhaps the main difference is the importance I accord to the naturalness principle as a basic intuitive criterion.

One might make the following summary of the situation: For formalism, consistency is the prime criterion. Feeling that consistency is not a sufficient guarantee or justification for the foundational nature of a system, even a comprehensive one, intuitionism seeks to build on an intuitive criterion of concreteness or constructibility. One might view my foundational systems as being properly contained in the former and not comparable with the latter. I would insist on consistency but would reject some consistent systems as unnatural. Though I regard many constructive notions as natural, I do not regard a notion as natural simply because, according to some metaphor of concreteness, it is more concrete than an alternative notion.

The term 'metaphor' is particularly appropriate here, I feel. Since everything in mathematics is abstract, and highly so, it does not seem to me that there is any immediately evident or *a priori* notion of constructibility which imposes itself upon us. Certainly it takes as great, if not greater, powers of abstraction to visualize many of the complicated constructive hierarchies currently in vogue than it does to visualize an infinite set such as that of the natural numbers or the real numbers. In fact, I feel that we may consider that one of the main philosophical problems in the epistemology of mathematics is that of finding the most natural notions of constructibility and of justifying the naturalness of certain current 'constructive' notions. I am thinking particularly of various versions of constructive analysis and of constructive hierarchies.

Université Laval, Quebec, Canada

NOTES

[1] This view would seem to bear some resemblance to ideas which Heyting attributes to Brouwer (cf. Kleene, 1952, p. 51). However, I do not draw the same conclusions as Brouwer.

[2] Interestingly enough, both Frege and Dedekind did attempt to deduce the axiom of infinity from the abstraction principle. A theorem of infinity is provable on the basis of abstraction and extensionality alone provided that the language is sufficiently rich (cf. Hatcher, 1968, pp. 103–9 and pp. 249–53).

BIBLIOGRAPHY

Carnap, R., 1970, 'The Logicist Foundations of Mathematics', in *Essays on Bertrand Russell* (ed. by E. D. Klemke), University of Illinois Press, Urbana, Ill.

Hatcher, W., 1968, *Foundations of Mathematics*, Saunders & Co., Philadelphia.
Kleene, S., 1952, *Introduction to Metamathematics*, N. J. Van Nostrand, Princeton.
Lawvere, F., 1969, 'Adjointness in Foundations', *Dialectica* 23, 281–96.
Mehlberg, S., 1962, 'The Present Situation in the Philosophy of Mathematics', in *Logic and Language*, D. Reidel, Dordrecht.
Morse, A., 1965, *A Theory of Sets*, Academic Press, New York.
Specker, E., 1962, 'Typical Ambiguity', *Logic, Methodology, and Philosophy of Science* (Proc. Int. Cong., 1960), Stanford, pp. 116–24.

CHARLES CASTONGUAY

NATURALISM IN MATHEMATICS
Comments on Hatcher's Paper

Hatcher's main theme (Hatcher, 1972), that a pragmatic approach to foundational problems is the most salutary one, is, I hope, nowadays generally agreed. I will examine instead his two other main proposals: that the most fruitful, or useful way to view mathematics is as the exact part of our thinking, and that naturalness is an important intuitive criterion in elaborating foundational systems. Both proposals are meant to be taken intuitively and pragmatically, that is, Hatcher gives no substantial explanation of what he means by 'exact' and 'natural', and gives no argument for the desirability of his proposals other than that they account, in his opinion, for certain facets of mathematical practice. I hold both proposals to be unacceptable, for the (not 'ultimate', but largely pragmatically inspired) reasons which follow.

I. EXACTNESS

While the philosopher is forever at grips with inexact concepts and with the fluctuating character of meaning, the mathematician tends to consider his conceptual universe as crystalline, and to regard his activity within it as a search for interrelations between its rigid parts. But in actual fact the concepts of mathematics are also subject to flux. Especially at the inception of a new way of perceiving some fragment of mathematics, some features of the fragment may yield to change – the Euclidean conception of geometry dissolves under a Cartesian approach, or the Leibnizian continuum is transfigured by Weierstrass' treatment of limits.

Even features which have been completely lost from sight can surprisingly reappear, as has the infinitesimal in non-standard analysis. But the conceptual context in which such features resurge can be so new as to pose serious problems of identity. This contextual flux affects even extremely 'constant' mathematical features: it can be asked, for example, whether the notion of natural number, treated in a category-theoretic setting, is the same as that envisaged by a constructivist such as Brouwer.

Or again, Cauchy's 'theorem', that the limit of a convergent sequence of continuous functions is continuous, though false in the Weierstrassian sense, is correct from a non-standard perspective (cf. Cleave, 1971).

And even when taken from a fixed viewpoint, a given mathematical concept – even such an apparently unproblematic one as that of a polyhedron (cf. Lakatos, 1963) – can often be successively analyzed through several degrees of precision, the result at each degree being considered as definitive until a new counterexample is hit upon. Not even in mathematics, then, is exactness absolute.

This continued evolution and critical reassessment of mathematical concepts inhibits, in my opinion, the construal of mathematics in such static terms as 'the exact part of our thinking'. It seems to me more desirable to take the dynamic aspect of mathematics as primordial, to view mathematics as essentially an activity, as part of the quest for exactness, and to regard as an instance of mathematics any more rigourous ordering, at a higher level of conceptual awareness, of phenomena apprehended at a lower level of intellectual organization.

It might be objected here that the difference between this view and Hatcher's resides only in the distinction between the activity and the finished product, and that both views are, therefore, equally tenable. I would reply that, perhaps more than anywhere else, in mathematics the activity eclipses the product: one *does* mathematics. It is much more important that a student of mathematics acquire certain reflexes of rigour and intuition than any given body of facts. When Hatcher himself elaborates on his metaphor, it is indeed in such active terms as 'objectifying', 'abstracting', and especially as a 'give and take between the conceptual and the formal, the intuitive and the logical'.

We might well, then, regard mathematics fundamentally as the modeling of the vague by the less vague. This in fact satisfies all that Hatcher requires of a characterization of mathematics: it consecrates the idea of the dual nature of mathematical thought (which, with all due respect to Lawvere, has been much more deeply elaborated in Gonseth's philosophy of science, cf. Gonseth, 1970), frees mathematics from any special source, or object, and obviously accounts for its non-numerical applications. Even more, this dynamic view points towards a solution to the problem of what significance to attribute to mathematics: mathematical concepts or theories draw their significance from the heuristic *milieux* which give rise to their formulation, from the more vague concepts or situations (physical or conceptual) which they purport to clarify.

I do not wish to further defend here the 'fruitfulness' of this move. Whatever its shortcomings, it improves the well-worn and uninformative (*Q*: What is exact? *A*: Mathematics) exact-part-of-thought definition, by explicitly acknowledging the openness of the notion of exactness, and by emphasizing its relative and local character. The move should appeal to category-theorists, in that it suggests we throw away the objects and just keep the morphisms, as they would put it. In matters of exactness, it is indeed the morphisms which chiefly count.

But the move would apparently not provide a 'useful' classification of a part of knowledge, in Hatcher's opinion. For he expresses a tendency to regard philosophical preoccupations (as fuzzy?) as distinct from mathematical ones, as if the levels of activity of philosophy and mathematics were entirely disjoint. Yet surely, the determinant characteristic of the exact philosopher is precisely his impatience with vagueness: "*Die Philosophie ist keine Lehre, sondern ein Tätigkeit ... Das Resultat der Philosophie sind nicht "philosphische Sätze", sondern das Klarwenden von Sätzen*" (Wittgenstein, 1922, 4.112). Or would my move make exact philosophy, which also indulges in the modeling of the vague by the less vague, not 'ultimate' enough to count as philosophy?

The interaction (or interference, if one likes) between philosophical and mathematical interests (and others) should be considered, at least potentially, to be more intimate than Hatcher's mathematism suggests. Far from being 'the most fruitful way to regard mathematics', Hatcher's view of mathematics closes too many doors.

II. NATURALNESS

As the exact philosopher also shares a deep distrust for universal and eternal foundational systems (cf. Bunge, 1962), objective criteria with which to gauge the excellence of a given (partial) foundational system are most welcome. But naturalness proves unacceptable to this end.

Etymologically, one would expect that a perfectly natural foundational system for some fragment of mathematics would be virtually indistinguishable from that fragment, in that it would imitate as closely as possible all, and perhaps *only* all, the features of the fragment. Such a foundational system would certainly enjoy coverage, that is, one would be able, within the system, to fully view, or represent, the fragment being founded. But how can such a system do more in fact than 'passively capture' the fragment?

From the context in which Hatcher first introduces the term, 'natural' seems indeed to imply, for him, 'no tension'. How then can a natural system 'lead and stimulate our intuition'? It seems clear, on the one hand, that an intuitively felt tension between concepts promises further development and clarification of these concepts, perhaps leading to final elucidation of the cause of tension, as exemplified by Specker's work on the abstraction principle. It seems equally clear, on the other hand, that no matter how thoroughly natural a foundational system is, such naturalness alone is no guarantee of its being essentially more suggestive than the body of intuitively conceived mathematics which it organizes. Suggestiveness would, I propose, more likely spring from a certain *un*naturalness, from a tension between distinct intuitions of a mathematical feature: Lawvere's category-theoretic formulation of number is suggestive precisely in its newness, in its departure from that conception of the integers hitherto accepted as natural, i.e. as counting sets. Such tension need not spring from any inconsistency between the distinct views.

Yet, from Hatcher's subsequent uses of the term, naturalness should, from all that I can see, also entail suggestiveness. I can only conclude that he expects too much of the one word, that it does not convey what is intended. Since in any case the use of such a realist terminology, with its possible beckoning to Platonism, runs the risk of awakening the ghosts of foundational schools 'fighting to kill one another' (Wang, 1958, p. 472), it would be more judicious to employ 'coverage' or 'adequacy' to cover the strict sense of 'naturalness'. Such terminology is philosophically neutral, and underlines more explicitly the relativity of the criterion to the foundational perspective taken. We might then find a foundational system 'good', in Hatcher's sense, if it is both adequate and suggestive – and of course consistent, as far as known. And we can leave 'natural' to be further overworked by category-theoretic structuralism.

III. ADEQUACY AND SUGGESTIVENESS

But rather than be satisfied with just switching words, let us try to gain a minimal insight into our two criteria. First, adequacy. The term evokes a certain selectiveness. A foundational system will be adequate for a given mathematical fragment *relative to a certain perspective taken*. Weierstrass, in his foundations for analysis, chose to ignore the infinitesimal; a category-theoretic formulation of a mathematical fragment will focus on certain relevant structural features of the fragment, and neglect other aspects. The foundational system must then provide an adequate setting for the study of those selected aspects of the fragment as viewed from the particular chosen perspective.

We now see how adequacy can point to more than just passive coverage, and blends into our second criterion, suggestiveness. For if the system is to be adequate in a deeper sense, it must not only provide the means to mirror the relevant logical structure, say, of the fragment, as seen from the given perspective, but should also provide a strong heuristic *milieu* which will suggest possible additional facts concerning the fragment, plausible hypotheses which were, nonetheless, less strongly suggested by the heuristics of the fragment. The heuristic power of the limit concept proved adequate, in this deeper sense, for the capture and development of the chosen aspects of analysis (continuity, differentiability, etc.), and category-theoretic formulations of parts of the mathematical corpus can, presumably, be similarly effective in describing and developing certain structural aspects of these parts.

A 'good' foundational system, then, should provide new reflexes of postulation with which to spur further investigation of the fragment. More than being just a language, it is important that a foundational system come equipped with its own vigorous heuristic component.

This clarifies the initially shocking claim which I made earlier, concerning the second criterion, suggestiveness: a 'good' foundational system is natural in its adequacy with respect to its logical and heuristic duties *vis-à-vis* the chosen perspective of the fragment, but is at the same time *unnatural* with respect to the fragment, in its neglect of certain aspects of the fragment, as well as in that its heuristic component must be markedly different from that of the fragment. This is why Hatcher's notion of 'naturalness', as criterion for a good foundation, causes discomfort and confusion. The adequacy-suggestiveness pair constitute a much more reasonable alternative.

The criterion of suggestiveness, of course, is largely post-applicable, and its premature usage can be little more than wishful thinking. Nevertheless I think it is worth retaining, not only to help dissipate the impression of mere passive coverage that 'adequacy' might give, but also to stimulate a healthy, Popperian attitude towards foundations, in the sense that we should not disdain to pursue the development of foundational tools which, at first, may strike us as unnatural. Nothing risked, nothing gained. And if crises arise, where theoretical hypotheses are found to be in open conflict with 'reality', far from being catastrophic, they lead to a better understanding of the model and of what is being modeled.

IV. FOUNDATION *v.* EXPLANATION

If we do want to bag the two birds (and possibly more) with one stone, returning to Hatcher's opening pages, we have the single notion which, without paradox, will do all that he wants, namely, explanatory value. Just by scratching the surface, it seems obvious that, in order to possess explanatory value, it is necessary that a foundational system at once enjoy coverage and improve our vision of the mathematical fragment in question; that is, that it be adequate and suggestive. We have seen how it is possible for a foundation to enjoy both.

This view of the fundamental role of a foundational system, as *explanans* for a part of mathematics, has been in fact proposed by Gödel, as reported with some emphasis by Mehlberg (1962, p. 86, Hatcher's reference). The passive, static conception of foundations as an unshakeable rock has given way to an active, dynamic conception, as the multi-faceted dance of explanation. Here, too, the objects are being abandoned in favour of the morphisms! A foundational system will be valued according to its heuristic power. Foundations has come a long way since Aristotle, and may well think of changing its name to something more in keeping with its modern role in mathematics.

Hatcher's naturalist terminology is not in tune with such a (thoroughly pragmatic) approach to foundations. To speak of natural and of unnatural foundations is to misrepresent the dynamism of mathematics, to mask its openness to change and diversification. To oppose a 'natural' (read: 'mathematical'?) intuition of the choice axiom to the 'logical' (read: 'unnatural'? 'unmathematical'?) intuition of the abstraction principle, is misleading, and simply less exact than to speak of the inadequacy of a foundational system which admits complete typical ambiguity, for a fragment of set theory which includes the choice axiom.

We will apparently have to live with a proliferation of set theories, providing as many divergent explanations of the set concept. Why tolerate, through an ill-chosen terminology, the possibility of insinuating a belligerent superiority of one system over another? Specker's result simply establishes that the logicist perspective on the set concept is different from that of the 'working mathematician'; which was quite plausible from the start, considering the gap between the heuristic *milieux* involved. There is nothing to be gained in talking past one another, as was done for so long in the controversies over the infinitesimal. Is nonstandard analysis to be rejected because it is too logicist in style? I only hope that category-theoretic messianism turns out to be as fruitful for foundations as logicism has been! Let us be (relatively) exact even in our choice of terminology for 'philosophical' discussions of foundations, and leave the doors open.

Hatcher's closing words illustrate, in fact, the possible connotation of ultimateness which I find most objectionable in his naturalness criterion. I do not expect that any philosopher would dare come up with an 'ultimate justification' of the naturalness of any constructivist notions. This is certainly one area where Hatcher's mathematism collapses, and where only very close collaboration between mathematician and philosopher (and psychologist, cf. Piaget's approach to constructivism by way of his genetic epistemology, in Beth and Piaget, 1961) can hope to shed some light. Nor would I expect the result to have any serious influence on mathematical practice. Absolutism is becoming hard to find in philosophy nowadays. Philosophers, too, are learning to de-emphasize the objects.

Univ. of Ottawa, Ottawa

BIBLIOGRAPHY

Beth, E. W. and Piaget, J., 1961, *Epistémologie mathématique et psychologie*, Presses Universitaires de France, Paris. English translation (1966): *Mathematical Epistemology and Psychology*, Gordon and Breach, N.Y.
Bunge, M., 1962, *Intuition and Science*, Prentice-Hall, Englewood Cliffs, N.J.
Cleave, J. P., 1971, 'Cauchy, Convergence and Continuity', *Brit. J. Phil. Sci.* **22**, 27–37.
Gonseth, F., 1970, 'Mon itinéraire philosophique', *Revue internationale de philosophie*, Nos. 93–94, Fasc. 3–4.

Hatcher, W. S., 1972, this volume, pp. 349–358.
Lakatos, I., 1963, 'Proofs and Refutations (I)-(IV)', *Brit. J. Phil. Sci.* **14** (1963–4), 1–25, 120–39, 221–45, 296–342.
Wang, H., 1958, 'Eighty Years of Foundational Studies', *Dialectica* **12**, 466–97.
Wittgenstein, L., 1922, *Tractatus Logico-Philosophicus*, Routledge and Kegan Paul, London.

PART V

PHILOSOPHY OF SCIENCE

RAIMO TUOMELA

DEDUCTIVE EXPLANATION OF SCIENTIFIC LAWS*

I

Since the appearance of Hempel's and Oppenheim's famous paper (1948) on the logic of scientific explanation a lively discussion has been going on concerning the logical criteria of adequacy for the explanation of singular events (states of affairs, processes, etc.). However, the theoretically much more important topic of the logic of the explanation of scientific laws has been neglected to a great extent in recent discussion. Notable exceptions to this claim are the works by Campbell (1920), Nagel (1961), Bunge (1967), and partly the recent articles by Ackermann (1965), Ackermann and Stenner (1966) and Omer (1970). Furthermore, in the debate between the representatives of the (or a) two-level picture of science (e.g. Hempel, Feigl) and the 'omnitheoreticians' (e.g. Feyerabend, Kuhn) the logical aspects of the explanation of (empirical) laws and theories by more developed theories has been treated to some extent. In any case the discussion on the explanation of laws has in general lacked the formal rigour and sophistication characteristic of the discussion of the explanation of singular events.

It has often been argued that the notion of (scientific) explanation is a pragmatic notion which does not have a clear-cut unambiguous objective formal structure. Sometimes these arguments purport to show that the relation of explanation holding between an explanans and an explanandum ceases to hold under the substitution of logically equivalent explanantia or explananda. Or then some other commonly accepted formal invariance conditions are claimed to fail (cf. p. 386). Often this type of argumentation is invoked to show the intensional (nontruthfunctional) character of the notion of explanation, which thus – *qua* intensional – cannot be clarified by an explicate (or explicates) within standard extensional logic. However, as we do not know of any convincing arguments to prove the latter we will proceed under the working hypothesis that for the time being it is meaningful to look for formal criteria that acceptable scientific explanations should satisfy.

On the other hand, it seems clear to us that the enormously difficult and complex notion of explanation will have a number of different objectified and idealized explicates – and not only one. This partly follows from the contextual characteristics of ordinary common sense explanations from which the notion or rather the various notions of scientific explanation have been obtained by the processes of abstraction, idealization, schematization, and objectification, and so forth. Common sense explanation seems to be dependent on the properties of the explainer and the explainee as well as on many other aspects of the context. All this is part of what if often meant by referring to explanation as a *pragmatic* notion. There are a number of pragmatic features which cannot be abstracted away in a full-blown theory of scientific explanation without too great a philosophical loss. Lacking a developed adequate pragmatics for the philosophy of science – which would be needed – something like Kuhnian paradigms presently seem to provide the best pragmatic background frame for a theory of scientific explanation.

Indeed, the general framework within which the problems in the present paper can be embedded is briefly this. Consider a fixed Kuhnian paradigm and within it a research programme incorporating a temporal sequence of theories $T_1, T_2, ..., T_n, T_{n+1}, ...$. We shall assume that the paradigm determines which (essentially general) statements qualify as scientific laws and which (sets of) statements represent acceptable theories. Furthermore, it will be assumed that the paradigm specifies the philosophical purposes of a scientific explanation. Until these conditions are fulfilled it does not seem meaningful to ask for a clarification of the logical and methodological features of scientific explanations.

Within the present frame one can start asking various questions pertaining to the growth of science as seen from the point of view of the present paradigm. For instance, what kind of philosophically, methodologically, and logically interesting transitions from some T_n to T_{n+1} can be found? In the present paper we shall consider only one special type of scientific growth occurring within this scheme, viz. that where T_{n+1} (deductively) explains and thus supersedes T_n. Here T_n can be an empirical law or theory and T_{n+1} a more comprehensive theory often introducing new explanatory ideas.

It has often been contended that in actual science a superseding theory T_{n+1} almost never deductively implies T_n. Rather T_{n+1} implies a theory or law T_n^* which can be considered an approximation of T_n and which is at least initially in experimental agreement with T_n. (The explanation of Kepler's laws by Newtonian mechanics is often mentioned as an example at this point.) Another way of putting this is to say that T_{n+1} explains T_n by correcting it at the same time. Undoubtedly this kind of correction process often takes place in science. Nevertheless, one may argue that this does not basically contradict the very idea of deductive explanation. For one may always argue that what one really aims at in this situation is an explanation and deeper understanding of the (extralinguistic) regularity allegedly described by T_n. However, in the process of explanation it is found that T_n^* rather than T_n seems to be a correct description of this regularity.

If this characterization of the situation is accepted there is nothing seriously wrong with deductive explanation, not at least on this point, even if deductive explanation alone does not then seem capable of accounting for transitions within actual scientific research programmes. (It should be remarked here that our notions of paradigm and research programme are to be understood in such a way that really revolutionary transitions always involve abandoning a paradigm and a research programme associated with it. Hence we do not have to take stand to questions of meaning variance, etc., which seem to involve a change of paradigm.)

Apart from some comments in the final section we shall in this paper concentrate on the logical and methodological rather than on the epistemological and metaphysical aspects of scientific explanation. Our starting point will be to regard explanation as an informative or information-providing argument. Unless considerably explicated and developed this idea is of course platitudinous. Explanation of laws can and has been regarded, for instance, as aiming at finding hidden causes for observable phenomena, or alternatively a deeper description of reality. Both of these basic types of explanation give us reasons for believing in the laws to be explained. At the same time it seems that an explanation (such as each of the above kinds) always carries a proper amount of relevant information concerning the explanandum and that this piece of information indeed constitutes the grounds for our believing in the explanandum. In this paper we shall mainly discuss information as measured by the logical (and especially quantificational) strength of statements. The sense or senses of information due to the introduction of new explanatory and perhaps ontologically more basic theoretical ideas will be emphasized, too.

II

1. Our main task in this paper is to examine some of the logical and methodological conditions of adequacy to be satisfied by deductive explanations of laws. Among other things we shall make an attempt to give a set (or actually several alternative sets) of necessary and sufficient logical conditions for an explanatory relation to hold between a theory and an empirical (or observational) generalization.

Among the attempts to give exact formal criteria of adequacy for deductive explanation, only those by Ackermann (1965), Ackermann and Stenner (1966), and Omer (1970) were also concerned with the explanation of general explananda in addition to singular ones. Ackermann's (1965) paper has been criticized and shown to be faulty in Ackermann and Stenner (1966). The latter paper again has been criticized by Omer (1970) and by Käsbauer (reported in Stegmüller, (1969). Omer also presented a model of explanation of his own based on general information-theoretic considerations.

We shall start by a concise description of Omer's model of explanation. It will be followed by a criticism. Finally we shall try to amend Omer's model in some respects and discuss the implications of our amended and enriched models. The final amended model to be presented is intended to apply to the explanation of laws only. Whether or not it can easily be applied to the explanation of singular explananda we shall not discuss here in any detail. Hence there is no need to go through the whole welter of counterexamples to various explicates of the deductive-nomological model of explanation, as these counterexamples have generally been restricted to the explanation of singular explananda.

Let us thus go on to a review of Omer's model of explanation with whose main ideas we in general sympathize. Omer starts by applying to scientific explanation one of Grice's principles concerning the informativeness of assertive discourse. This gives the first general criterion for explanation (see Omer, 1970, p. 419):

R_I In the explanans no sentence which is less informative in the topic should be given when it is possible to give a more informative one.

Next Omer states as 'an important subcase' of this general requirement:

R_{Is} No sentence in the explanans should be of less informative content than the explanandum.

The explanandum-sentence is considered by Omer to be a true statement 'on the topic'. R_{Is} then indeed follows from R_I with the additional premise that it is possible to give an explanans more informative than the explanandum.

One important question left open by the above requirements is how to measure the information content of a sentence. Omer uses essentially the logical content explicate due to Carnap's theory of semantic information. According to it, the content of a statement is identified with its logical strength (and formally measured as one minus the logical probability of the statement). However, following Omer, we do not below utilize the metrical cont-measure at all. The only information-theoretic notion concretely needed in Omer's account is that of noncomparability of statements. We shall say that two statements.

> p and q are *noncomparable* with respect to their information content if and only if
> not $\vdash p \supset q$ and not $\vdash q \supset p$.

It is important to notice what is involved in this notion of noncomparability. Two statements are noncomparable exactly when each of them includes some content-elements not included in the other one. Thus it is quite possible that they contain common content-elements. Indeed, unless this were the case within deductive explanation, the requirements concerning the noncomparability of the elements of the explanans and the explanandum to be considered below could not be reconciled with the idea of the explanans logically implying the explanandum. (Consider e.g. Craig's interpolation theorem from which it follows that there has to be some amount of common content between two statements one of which implies the other.)

As a side remark it can be pointed out here that within nondeductive explanation, measures of transmitted information (the inf-, cont- and entropy-measures) have been used to measure the goodness of an explanation. The more an explanans transmits or conveys information concerning the explanandum, the better is the explanation. However, in this sense all deductive explanations are maximally good. For if the explanans logically implies the explanandum, our coming to know the explanans completely relieves us of our agnosticism or uncertainty (these notions understood in the sense of the semantic theory of information) concerning the explanandum. Thus within deductive explanation this kind of information-theoretic explicate of explanation is of no help.

Let us now go back to Omer's general requirements for deductive explanation. In terms of logical content we now have (Omer, 1970, p. 422):

R'_I In the explanans no sentence in the topic which is of less logical content should be given when it is possible to give a sentence with more logical content.

R'_{Is} No sentence in the explanans should be of less logical content than the explanandum.

The content of the requirement R'_I (over and above that of R'_{Is}) can be illustrated by the following example. Suppose the statements $p \vee r$ and $p \vee -r$ are proposed to be included in an explanans. Now jointly they are equivalent to p but both of them are individually of less logical content than p as p implies them. Thus, according to R'_I instead of $p \vee r$ and $p \vee -r$ we should use p in the explanans.

Next consider the following example where T is a theory and L a law:

$$\frac{T \quad T \supset L}{L}$$

(Notice that here the second premise is logically equivalent with $-T \vee L$, wich clearly follows logically from L.) The requirement R'_{Is} clearly blocks, among other things, this kind of trivial deductions from being explanations. But it does not block the following paradoxical case given in Hempel and Oppenheim (1948) p. 273):

$$\frac{B \quad K}{K}$$

In this self-explanation K could stand for Kepler's laws and B for Boyle's law. Apparently R'_{Is} has to be strengthened. Omer also comes to this conclusion, although on the basis of examples with singular explananda. Omer now proposes (Omer, 1970, p. 423):

R''_{Is} The explanandum should be noncomparable with any of the sentences of the explanans.

But what is meant by a sentence here? Is the conjunction of all the sentences occurring in the explanans a sentence? If so, then certainly R''_{Is} is not acceptable as it is clearly in blatant contradiction with the basic requirement that the explanans should logically imply the explanandum. Apparently some clarification is needed. This may be what Omer thinks, too, as he proceeds to give a more technical equivalent to R''_{Is}, even if he does not motivate his move.

Before giving the final version of the noncomparability requirement we shall have to introduce some technical apparatus. Let us consider some scientific language $\mathscr{L}$ with specified logical and extralogical vocabulary, well formed formulas, logical axioms and rules of inference. In the manner of Ackermann and Stenner (1966, p. 169) we now define the notions of a sequence of truth functional components and a set of ultimate sentential conjuncts of a given formula (sentence) T.

A sequence of statemental well formed formulas $\langle W_1, W_2, ..., W_n \rangle$ of $\mathscr{L}$ is a *sequence of truth functional components* of T if and only if T may be built up from the sequence by the formation rules of $\mathscr{L}$, such that each member of the sequence is used exactly once in the application of the rules in question. The W_i's are thus to be construed as tokens. Thus, for instance, $\langle (x) F(x), (x) F(x) \rangle$ and not $\langle (x) F(x) \rangle$ is a sequence of truth functional components of the formula $(x) F(x)$ & $(x) F(x)$. The formation rules of $\mathscr{L}$ naturally have to be specified in order to see the exact meaning of the notion of a sequence of truth functional components of a theory finitely axiomatized by a sentence T.

A *set of ultimate sentential conjuncts* Tc of a sentence T is any set whose members are the well formed formulas of a longest sequence $\langle W_1, W_2, ..., W_n \rangle$ of truth functional components of T such that T and W_1 & W_2 & ... & W_n are logically equivalent. If T is a set of sentences then the set Tc of ultimate conjuncts of T is the union of the sets of ultimate sentential conjuncts of each member of T. We may notice here that although by definition the Tc-sets of two logically equivalent theories are logically equivalent they need not be the same.

Now we are ready to state the final version of the noncomparability requirement for a Tc of a theory T constituting an explanans (cf. Omer, 1970, p. 424):

R_{Is}''' For any Tc_i in Tc, Tc_i is noncomparable with the explanandum.

The requirement R_{Is}''' seems to avoid at least the paradoxical examples of self-explanation concerned with singular explananda to be found in the literature. Furthermore, it seems to do the same in the case of the explanation of laws as well (cf. the example by Hempel and Oppenheim given above).

Next we shall state some necessary conditions for an explicate of deductive scientific explanation which also Omer accepts (although in a somewhat different formulation). Let us write E(L, T) for 'L is potentially explained by T'. Here T and L are statements of our scientific language $\mathscr{L}$. (Sometimes we shall also denote by T and L sets of statements, which does not make any essential difference as long as these sets are finite.) As above, Tc is a set of ultimate sentential conjuncts of T. For simplicity, in the sequel we shall be concerned with E(L, Tc) rather than E(L, T). (Recall that Tc and T are logically equivalent. When speaking of E (L, Tc) Tc is to be understood as the conjunction of its members rather than a set, which makes no essential difference here.) Now we are ready to give some conditions for an explanatory relation E(L, Tc) to hold between Tc and L:

E(L, Tc) satisfies the logical criteria of adequacy for deductive explanation only if

(1) $\{L, Tc\}$ is consistent;

(2) $Tc \vdash L$;

(3)   some Tc_i in Tc is a universal law;

(4)   for any Tc_i in Tc, Tc_i is noncomparable with L.

In view of what was said in Section I of this paper, (1) seems acceptable. Requirement (2) is obvious. As to (3) we shall here require of a law only the syntactic feature that it essentially contains at least one universal quantifier, and no fuller treatment of lawlikeness, beyond giving this necessary condition, is attempted at in this paper. Requirement (4) is just R_{Is}'''. Notice that in this model L can be either a singular or a general statement.

Can the conjunction of the conditions (1)–(4) be regarded as a sufficient condition of explanation as well? It seems that the resulting model of explanation is too wide. For if we have found an explanans for a statement, say $G(a)$, within this model, the same explanans will also be allowed to be an explanans for all the disjunctions in which $G(a)$ occurs as a disjunct and which do not contradict condition (4). In accordance with Omer we are here inclined to consider this consequence unacceptable (or at least undesirable).

How can we exclude such counterintuitive explanations? Let us first consider Omer's own argumentation. (See Omer, 1970, p. 425) First he argues that the phrase 'in the topic' of the original R_I secures that all additional pieces of information are relevant information. This seems acceptable, though vague. But then he claims that R_I is to be interpreted or qualified so as to rule out all redundant information (i.e. redundant over and above what is needed for T to imply L). This requirement, however, is in clear contradiction with R_I' (supposedly equivalent with R_I). Indeed, as R_I clearly seems to call for maximally informative explanantia, Omer on the contrary is here led to look – *ceteris paribus* – for minimally informative explanantia. This is seen from the following, which – apart from an index correction – is Omer's last condition (Omer, 1970), p. 426):

(5) It is not possible to find sentences $S_i, ..., S_r$
$(r \geqslant 1)$ such that for some $Tc_j, ..., Tc_n (n \geqslant 1)$:
$Tc_j \& ... \& Tc_n \vdash S_i \& ... \& S_r$
not $S_i \& ... \& S_r \vdash Tc_j \& ... \& Tc_n$
$Tc_s \vdash L$
where Tc_s is the result of the replacement of
$Tc_j, ..., Tc_n$ by $S_i, ..., S_r$ in Tc.

Omer now claims that the conditions (1)–(5) are necessary and sufficient for deductive explanation.

Condition (5) requires that within Omer's model of explanation we choose as the explanans a minimally strong set of statements. To explain something is then to give a specific (minimal) amount of (proper) information. Omer claims with examples that his model is free from the paradoxes found against the previous models of deductive explanation. (See Omer, 1970, pp. 422f., 427–432 for the discussion of the arguments by Hempel and Oppenheim, 1948; Eberle *et al.*, 1961; Kaplan, 1961; Kim 1963; Ackermann, 1965; Ackermann and Stenner, 1966.) We shall not comment upon them here directly, partly because the paradoxes deal only with singular explananda. Furthermore, we shall below show that there are so obvious flaws in Omer's model that they lead to the rejection of it in its present form.

2. Let us now proceed to a criticism of Omer's model which will lead us to two amended versions.

Let us consider the following paradigm case for the explanation of singular explananda:

$$(x)(F(x) \supset G(x))$$
$$\underline{F(a)}$$
$$G(a)$$

Does it satisfy Omer's conditions? Somewhat it surprisingly it does not, for it does not satisfy condition (5). Contrary to Omer's claim that conformity to the above condition (5) assures the relevance of the law more than ever, our explanans would here become simply $\{G(a)\}$. But this explanans obviously contradicts both condition (3) and (4). In other words, Omer's model of explanation is clearly inconsistent. An apparent and trivial flaw in Omer's model seems to be this. A requirement to the effect that the statements $S_i, ..., S_r$ should be (universal) laws whenever the statements $Tc_j, ..., Tc_n$ are, seems to be missing from condition (5). But this correction does not yet suffice to establish a consistent model of explanation. For the amended condition (5) will still always lead to a contradiction in those cases where the explanandum is a law statement. That this is the case is seen by taking simply the explanandum as the only S_i. This makes the amended condition (5) true but it falsifies (4). To block this we have to insert a conditional consistency requirement. Thus we propose the following amendation of (5):

(5')   It is not possible without contradicting any of the previous conditions to find sentences $S_i, ..., S_r$ ($r \geq 1$) at least some of which are essentially universal such that for some
$Tc_j, ... Tc_n$ ($n \geq 1$):
$Tc_j \& ... \& Tc_n \vdash S_i \& ... \& S_r$
not $S_i \& ... \& S_r \vdash Tc_j \& ... \& Tc_n$
$Tc_s \vdash L$
where Tc_s is the result of the replacement of $Tc_j, ..., Tc_n$ by $S_i, ..., S_r$ in Tc.

The model satisfying (1), (2), (3), (4), and (5') (call it Omer's corrected model) really seems to be a 'minimal law' model according to which an explanation provides a minimal amount of specific information. To illustrate, if we have a singular explanandum $G(a) \vee H(a)$ and the initial condition statement $F(a)$, then we have to choose in the explanans the law $(x)(F(x) \supset G(x) \vee H(x))$ and not, say, $(x)(F(x) \supset F(x))$. The latter law is stronger than the former, and it, together with $F(a)$, suffices to imply $G(a) \vee H(a)$. This model also seems to be a possible candidate for the deductive explanation of scientific laws.

As we noticed, neither the model (1)–(5) nor the present version (1)–(5') satisfies the general information theoretic principle R'_I; and they satisfy the original version R_I only if it is interpreted in a rather queer way which is indeed the opposite of our (and Omer's) original reading of it. We suggest that the present model should rather be considered as satisfying something like the following strong modification of R_I (actually compatible with what Omer says on p. 425 of his paper):

R^*_I The explanans should contain redundant information only if is it *relevant*.

The notions of information and redundant information are to be understood in the previous information theoretic sense in which information content and logical strength go together. The problem left then is what is to be understood by relevancy. One possibility is that formalized by (5'). But it is not the only possible explicate.

Consider the following two proposed explanations for the same singular explanandum:

$$
(*)\quad \frac{\begin{array}{l}(x)(y)(F(x,y)\supset G(x,y))\\ F(a,b)\end{array}}{G(a,b)}\quad \begin{array}{l}\text{Tc}_1^*\\ \text{Tc}_2^*\end{array}
$$

$$
(**)\quad \frac{\begin{array}{l}(y)(F(a,y)\supset G(a,y))\\ F(a,b)\end{array}}{G(a,b)}\quad \begin{array}{l}\text{Tc}_1^{**}\\ \text{Tc}_2^{**}\end{array}
$$

The more general proposal (*) is not accepted as an explanation by Omer's corrected model, but (**) is. Now one may argue that there is a sense of explanation in which (*) counts as an explanation of $G(a, b)$, perhaps even as a much better one than (**). In this sense of explanation the information provided by the additional quantificational generality of the law Tc_1^* over and above the generality of the law Tc_1^{**} may be considered *relevant* redundant information (in the sense of R_1^*). (Cf. e.g. Hintikka, 1970, for an argument concerning the increase of 'surface' information with quantificational depth, and cf. Hintikka and Tuomela, 1970, for an argument concerning how observational information content increases with quantificational depth due to the introduction of new explanatory concepts.)

Let us thus accept that there is a sense of information in which the increase in quantificational generality provides relevant information. How can this observation be taken into account in a more technical fashion? We propose the following. Let us write $p \vdash_p q$ when q is deducible from p by using sentential logic only. Now (5') is to be replaced by:

(5'') It is not possible, without contradicting any of the previous conditions for explanation, to find sentences $S_i, ..., S_r$ $(r \geqslant 1)$ at least some of which are essentially universal such that for some $\text{Tc}_j, ..., \text{Tc}_n$ $(n \geqslant 1)$:
$\text{Tc}_j \& ... \& \text{Tc}_n \vdash_p S_i \& ... \& S_r$
not $S_i \& ... \& S_r \vdash \text{Tc}_j \& ... \& \text{Tc}_n$
$\text{Tc}_s \vdash L$
where Tc_s is the result of the replacement of $\text{Tc}_j, ..., \text{Tc}_n$ by $S_i, ..., S_r$ in Tc.

The resulting model of deductive explanation thus explicitly defines the relation E(L, Tc) by the conditions (1), (2), (3), (4), and (5″). This model accepts both (*) and (**) as explanations. In principle the explananda L can again be either singular or general. To get a simple example with a general explanandum we can change (**) by adding a third argument to the two-place predicates F and G and by universally quantifying over the new argument (say z). The resulting explanation (call it (″)) qualifies in our present model but not in Omer's corrected model. The same holds true for the analogue (′) of (*). The only explanatory argument which in the present frame would be accepted both by our present model and Omer's corrected model is one in which only z is quantified. (Call this explanation (‴).)

We propose that the conditions (1)–(5″) are to be accepted as necessary and sufficient for determining the logical structure of the explanation of (at least) *singular* explananda. (Below we shall still require something more of good explanations of scientific *laws*.) It should be noticed that the conditions (1)–(5″) are not restricted to any specified formal language, as most of the previous models of explanation have been. Therefore it should be widely applicable. In addition, we have not made any philosophical restrictions as to the substantial content of explanations, but have restricted our analysis to the purely formal aspects of explanation.

3. At this point we shall recall one of our claims from Section I of this paper. There we mentioned that particularly in the explanation of laws one introduces theories which contain new explanatory concepts. To make justice to this we shall assume that we are dealing with an interpreted scientific language whose extra-logical vocabulary is dichotomized into a set λ of observational (or descriptive, or old) predicates and a set μ of theoretical (or explanatory, or new) predicates. The only thing we essentially require of this dichotomy here is that the members of μ cannot in *all* contexts be used in direct reporting. Therefore they have to be connected by correspondence rules to the members of λ, at least in some contexts, to be methodologically useful (and perhaps they are not even fully intelligible without that). We shall below assume that our explanandum laws contain only members of λ, and that the explanantia always contain at least some formulas in which members of μ occur.

In the explanation of a law $L(\lambda)$ the explanans will be a theory axiomatized by a sentence $T(\lambda \cup \mu)$ (or, alternatively, a $Tc(\lambda \cup \mu)$ corresponding to $T(\lambda \cup \mu)$). One can now distinguish the following three parts in the theory:

(1) the Campbellian part (or core theory) which consists of those statements of $T(\lambda \cup \mu)$ which are solely in the vocabulary μ;

(2) the observational content of $T(\lambda \cup \mu)$ which consists of those statements of $T(\lambda \cup \mu)$ which contain members of λ only;

(3) the correspondence rules of $T(\lambda \cup \mu)$ which consists of all the other statements of $T(\lambda \cup \mu)$ and are in the vocabulary $\lambda \cup \mu$.

Let us call the deductively closed sets comprising the components (1), (2), and (3) by $H(\mu)$, $O(\lambda)$, and $C(\lambda \cup \mu)$, respectively. For first-order languages, the decomposition of a sentence $T(\lambda \cup \mu)$ can be accomplished by means of, for instance, the reduction operation defined in Hintikka and Tuomela (1970). More generally, Craig's general replacement method shows how this can be done for deductive systems in the case of almost any formal system (cf. Craig, 1956). (We shall below denote by $Cn(T(\lambda \cup \mu))$ ($=H(\mu) \cup O(\lambda) \cup C(\lambda \cup \mu)$) the deductive closure of $T(\lambda \cup \mu)$, and by the Craigian μ-reduct of a deductive system we shall mean the recursively axiomatizable part of $Cn(T\lambda \cup \mu))$ solely in the vocabulary μ.)

When we below investigate how a theory $T(\lambda \cup \mu)$ explains a law $L(\lambda)$ this problem can thus also be considered as how a core theory $H(\mu)$ conjoined with some suitable correspondence rules $C(\lambda \cup \mu)$ explains a law $L(\lambda)$ belonging to $O(\lambda)$.

As we have repeatedly pointed out, there are many ways to explicate the notion of explanation. Furthermore, one may argue about which properties of the relation $E(L, Tc)$ are analytic and which are synthetic. We shall not take a definite standpoint to this here but, rather, we shall liberally try to take into account all the logically and methodologically important features of $E(L, Tc)$.

In the model of explanation satisfying (1), (2), (3), (4), and (5″) a number of important aspects of explanation (e.g. proper informativeness, prohibition of completely circular explanation, etc.) are already incorporated. But there are still at least two logically explicable properties to be considered (nonlogical ones will be discussed later). These are the requirements of observational creativity of the explanans over the explanandum law and the theoretical noncreativity of the explanans with respect to its Campbellian part.

As to the first of these let us quote Nagel. "At least one of the premises in the explanation of a given law will meet the following requirements: when conjoined with suitable additional assumptions the premise should be capable of explaining other laws than the given one; on the other hand it should not in turn be possible to explain the premise with the help of the given law even when those additional assumptions are adjoined to the law" (Nagel, 1961), p. 36). We shall accept a strengthened form of this condition, and its formalized version will appear below as condition (6) among our additional desiderata for explanation.

The theoretical noncreativity condition has been formulated by Nagel as follows: "the introduction of new correspondence rules does not change either the formal structure or the intended meaning of the theory, though new rules may enlarge the theory's range of application" (Nagel, 1961, p. 102. Also see Campbell, 1920, p. 133). In Hintikka (1972) this condition was dubbed the theoretical adhockery condition, as it requires that no *ad hoc* theoretical principles, but only (or at most) new correspondence rules, should be added when applying the theory to explain new laws. However, this condition cannot be applied unqualified to all explanation as clearly the core theory $H(\mu)$ must be allowed to grow if this growth takes place 'in a fruitful way'. We shall not include the theoretical noncreativity (or adhockery) requirement among the conditions of our model proper but as a separate condition ((7) below) to be applied when the above qualification can be accepted. (After this accounting for this requirement, all of Nagel's logical conditions of adequacy for the explanation of laws are taken case of by our conditions; cf. Nagel, 1961, pp. 33 ff, and p. 102.)

Now our final model of explanation of scientific laws can be stated in full as follows. As before, let thus T be a statement. Tc a set of ultimate sentential components of T (or actually a conjunction of components in the context E(L, Tc)), and L a statement to be explained. Then we say that the relation E(L, Tc) satisfies the *logical* conditions of adequacy for the deductive explanation of scientific laws if and only if

(1) $\{L, Tc\}$ is consistent;
(2) $Tc \vdash L$;
(3') Tc contains universal formulas at least some of which contain members of μ and L contains only general statements in λ;

(4)   for any Tc_i in Tc, Tc_i is noncomparable with L;
(5″) it is not possible, without contradicting any of the previous conditions for explanation, to find sentences $S_i, ..., S_r$ $(r \geqslant 1)$ at least some of which are essentially universal such that for some $Tc_j, ..., Tc_n$ $(n \geqslant 1)$:
$Tc_j \& ... \& Tc_n \vdash S_i \& ... \& S_r$
not $S_i \& ... \& S_r \vdash Tc_j \& ... \& Tc_n$
$Tc_s \vdash L$
where Tc_s is the result of the replacement of $Tc_j, ..., Tc_n$ by $S_i, ..., S_r$ in Tc.

As an additional desideratum we consider:

(6)   If $Cn(Tc) = H(\mu) \cup O(\lambda) \cup C(\lambda \cup \mu)$, then there is a $L' \neq L (L \in O(\lambda), L' \in O'(\lambda))$ such that not $\vdash L \supset L'$ and such that $E(L', Tc')$ for some Tc' (of a T') for which $Cn(Tc') = = H'(\mu) \cup O'(\lambda) \cup C'(\lambda \cup \mu)$ and $H \subseteq H'$; in addition, not $\{L\} \cup (H' - H) \cup (O' - O) \cup (C' - C) \vdash H$.

(I owe Prof. W. Stegmüller my present formulation of the last part of condition (6).) The optional theoretical adhockery condition, to be applied if no fruitful growth is present, is:

(7)   If $H(\mu)$ is an explanatory-core theory, and if, for some L, $E(L, Tc)$ with $Cn(Tc) = H_r(\mu) \cup O(\lambda) \cup C(\lambda \cup \mu)$, then the difference $H - H_r = \emptyset$. (H_r is the Craigian μ-reduct of $Cn(Tc)$).

It can be noticed that none of the above conditions is restricted to a specific formal system (such as first-order predicate calculus). Therefore they should have a wide area of application.

Let us call a model of deductive explanation of laws for which (1), (2), (3'), (4), and (5″) hold the *DEL-model*. If it is not required that theoretical concepts be essential for deductive explanation our old (3) can be substituted for (3') in the above model (and the appropriate notational changes can be made elsewhere) and we get the model discussed in subsection II.2. Call it the *weak DEL-model*. (An analogous remark holds for the situation where one is not willing to dichotomize the extralogical concepts in the manner assumed here.)

Earlier we described three simple law-explanations ('), (″), and (‴)

which all qualified as explanations within the weak DEL-model. If, in addition, the three-place predicate F is theoretical, all these explanations are accepted by the DEL-model, too. Explanations (') and ('') carry more relevant and deeper (quantificational) information than ('''). One consequence of this is that they in addition satisfy (6) which (''') fails to satisfy. (Notice that the fate of condition (7) cannot be determined before fixing an initial core theory $H(\mu)$.) As an additional example, consider the following simple chain explanation:

$$\frac{(x)(G_1(x) \supset F(x))}{(x)(F(x) \supset G_2(x))}$$
$$(x)(G_1(x) \supset G_2(x))$$

where $\lambda = \{G_1, G_2\}$ and $\mu = \{F\}$. Here G_1 and G_2 might be observable symptoms of a virus F.) This explanation is accepted by the DEL-model. But its shallowness is reflected by the fact that it does not satisfy (6).

As a scientific illustration of the DEL-model we can consider, for instance, the explanation of the well-known probability matching law. This law is explained by the linear learning theory of Bush and Mosteller. But it is also explainable by the general stimulus sampling theory of Suppes and Estes which in turn explains the linear learning theory as its special case.

Let us next state some formal properties of E(L, Tc) in the DEL-model. Most of them will also apply to the weak DEL-model, as can easily be seen.

First, it is easy to verify that in the DEL-model

(a)   E(L, Tc) is asymmetric
(b)   E(L, Tc) is irreflexive
(c)   E(L, Tc) is not transitive.

In the weak DEL-model, E(L, Tc) is easily seen to be irreflexive and neither symmetric nor transitive. To see that it is not transitive assume that E(L, Tc) and E(Tc, Tc') for some L, Tc, Tc'. Then we may possibly have $\vdash Tc'_j \supset Tc_j$, though not $\vdash Tc_j \supset L$, for some $Tc'_i \in Tc'$, $Tc_j \in Tc$. Thus even if not $\vdash Tc_j \supset L$ it may happen that $\vdash Tc'_i \supset L$. Hence condition (4) fails to hold. But still E(L, Tc) is not intransitive in the weak DEL-model.

Next we observe that:

(d) If $E(L, Tc)$ and if for some Tc', $\vdash Tc' \supset Tc$ (assuming not $\vdash_p Tc' \supset Tc$), then $E(L, Tc')$, provided every $Tc'_i \in Tc'$ is non-comparable with L.

(e) If $E(L, Tc)$ and if for some T' such that $\vdash T \equiv T'$, T and T' possess identical sets of ultimate sentential components (i.e. $Tc \subseteq Tc'$), then $E(L, Tc')$.

(f) If $E(L, Tc)$ and for some L, $\vdash L \equiv L'$, then $E(L', Tc)$, provided that, for all Tc_i in Tc, Tc_i and L' are noncomparable.

It should be noticed, however, that a DEL-explanation is not invariant with respect to the substitution of logically equivalent explanantia or explananda. This lack of linguistic invariance may be taken to reflect the *pragmatic* nature of our notion of explanation. How you *state* your explanatory argument makes a difference.

For some purposes it may be adequate to define a notion of indirect explanation. We may say that if $\vdash T \equiv T'$, $E(L, Tc)$ for some Tc of T and $Tc \neq Tc'$ (for all Tc'-sets of T'), then T' explains L *indirectly*.

The properties (d), (e), (f) and (g) below hold for the weak DEL-model as well.

It is easily seen that the explanatory relations of the DEL-model and of the weak DEL-model do *not* satisfy the following debated subset and conjunction properties:

If $E(L, Tc)$ and L' is a conjunct of L, then $E(L', Tc)$.
(Subset property)
If $E(L_i, Tc)$ for all L_i of an $L = L_1 \& \ldots \& L_i \& \ldots \& L_n$, then $E(L, Tc)$. (Conjunction property)

This means that the explanandum can be neither weakened nor strengthened without the explanation possibly ceasing to be valid.

We have already noticed that in the DEL-model a law L may have explanantia differing in their quantificational strength (and information content). The DEL-model does require 'minimal' explanation (in Omer's sense) only within explanans-candidates of the same quantificational strength. Thus we have a tree-structured hierarchy of explanations such that on each level of this quantificational hierarchy the minimality principle is satisfied. We can finally notice that the branches of the explanation-tree for an L may be created in diverse logically imcompatible ways:

(g) If $E(L, Tc)$ and $E(L, Tc')$, then it is possible that Tc and Tc' (and hence the corresponding theories T and T') are mutually logically incompatible.

The explanantia in one and the same branch are of course logically compatible, as the deeper explanantia quantification-theoretically imply the shallower explanantia for L.

III

We shall conclude this paper by making a few sketchy philosophical and methodological comments concerning the above results and by briefly discussing a few ideal nonlogical conditions of adequacy for the deductive explanation of scientific laws.[1] The various definitions for deductive explanation given in the previous section were only concerned with *potential* explanation. To make a potential explanation *actual* some requirements concerning the truth and/or the evidential support of the explanandum and its explanans are needed. One plausible candidate, which we shall adopt, is simply:

(8) If $E(L, Tc)$ then L and Tc should be accepted as true.

The term 'true' of course means something like 'true in the actual world' or 'true in intended model(s)'. The notion of acceptance is a notoriously tricky one, and here we shall not attempt to give an analysis of it. (It may be the case that most actually accepted scientific theories are false and even known to be false, but this does not speak against accepting (8) as an ideal for good explanatory theories.)

There is, however, one familiar requirement related to how we accept our explanans as true that we want to make more explicit. This is a requirement related to condition (6) and prohibiting what might be called 'confirmational adhockery' (cf. Nagel (1961), p. 43). We shall adopt the following weak version of it for our DEL-model:

(9) If $Cn(Tc(\lambda \cup \mu)) = H(\mu) \cup O(\lambda) \cup C(\lambda \cup \mu)$ and if $E(L, Tc)$, then $H(\mu)$ is – through some set $C'(\lambda \cup \mu)$ of correspondence rules – adequately supported by evidence based on data other than the observational data upon which the acceptance of the explanandum law L is based.

(Notice that (9) still allows that $O(\lambda) = \{L\}$, although it prohibits that $O'(\lambda) = \{L\}$.) The conditions (8) and (9) are necessary conditions to make the DEL-model for potential deductive explanation a model for actual (in the above sense) deductive explanation as well.

As we have seen, the DEL-model (together with conditions (6) and (7)) seems to be capable of giving an adequate account of many troublesome issues in the theory of explanation while it correctly emphasizes some features considered essential for good deductive explanations of scientific laws.

Based on our discussion in Section I we found condition (1) – the consistency condition – of the DEL-model acceptable. Condition (2) is obviously also acceptable. Condition (3') will be commented upon below. Conditions (4) and (5") were arrived at from general information-theoretic considerations. (4) was found to take care of circular explanations of the most vicious kind. (5") is supposed to guarantee that an explanation provides a proper amount of relevant (information-theoretically) redundant information. Condition (6) refers to the dynamics of theorizing. According to (6), it is essential that an explanatory theory be at least capable of growth. Condition (9) then actually realizes this possibility by requiring the growth of theory in the evidential sense. It can be noticed here that our condition (6) guarantes against *ad hoc$_1$*-type and condition (9) against *ad hoc$_2$*-type of the three kinds of *ad hoc*-explanations in Lakatos (1970). In addition, our theoretical adhockery condition (7) seems to be a special kind of Lakatos' *ad hoc$_3$*-explanations.

Condition (3') serves to indicate our emphasis on the importance of new theoretical ideas (concepts). As Campbell has frequently emphasized, an explanatory theory shall 'add to our ideas and the ideas which it adds shall be acceptable' (Campbell (1921), p. 83). Even if our condition (3') as such says nothing about the difficult and debated notion of 'acceptability' of the concepts in μ, still it – and the DEL-model as a whole – refers to *interpretative* rather than *phenomenological* theorizing (cf. Bunge, 1968). The explanation of observational laws by theories introducing new ideas (hidden causes, etc.) is often regarded as ontologically and epistemologically deeper, and it has been considered to add to our understanding of the world more than mere phenomenological explanation (explaining laws with the help of theories where no new kind of concepts occur). As we noticed, this kind of penetration into deeper levels of reality was in part and in addition made possible by the hierarchical organization (tree-structure) of explanations accomplished by condition (5″). (In a model of explanation based only on minimal theory explanations – such as Omer's corrected model – this would not have been possible.)

The additional and optional desideratum – our condition (7) – was introduced to avoid theoretical adhockery. This condition was above also called the theoretical noncreativeness condition as it requires that new correspondence rules added to the core theory to make it explain some new observational laws should be noncreative with respect to their λ-content. Hence one way to satisfy this condition is to give noncreative definitions of the observational concepts in terms of the theoretical concepts of the full theory. For instance, explicit definitions of the observables in terms of the theoreticals – rather than *vice versa* – would be such definitions (cf. Ramsey, 1931; Sellars, 1961; Carnap, 1966; and Hintikka 1972). In this case one can even theoretically eliminate observational or λ-concepts in terms of theoretical ones. Theoretical explanation (such as reductive explanation) seen as redescription or reinterpretation can be considered to require ultimately just this. Applied to intertheoretic reductive explanation this is seen to imply that every case of 'principle-reduction' can be performed at least as well by accomplishing a 'concept-reduction' simultaneously (see Hempel, 1969, for these notions).

Interpretative theorizing thus seems ideally to require explicit definitions of observables in terms of theoreticals while phenomenological theorizing seems to imply the reverse at least asymptotically.

Nagel has required that in the explanation of laws the explaining theory should be more *general* than the explanandum law (Nagel, 1961, pp. 37–8). As the generality in question is denied to be proportional to logical strength, the following suggestion seems to be at hand. A sufficient, though perhaps not necessary, condition for arriving at Nagel's generality would be to define explicitly the concepts occurring in the law in terms of theoreticals such that each definition is a conjunction of at least two theoreticals. For then one can in an obvious sense say that observable concepts are applications of theoretical concepts and that every theoretical concept is more general and broader in its scope than each observational concept in whose definiens it occurs.

Our final remark is concerned with the relation between explanation and various notions of inductive support. It seems that between such notions as explanation, generalization, confirmation, corroboration, etc., there are many connected features. Indeed, it might be required that theories of explanation and inductive support should always be developed simultaneously. One argument for this arises from an investigation by Smokler (1968). He pointed out that there are two essentially different notions of inductive support, one corresponding to 'abductive inference' and the other to 'enumerative induction') the latter includes eliminative induction). Within abductive inference it is true to say that evidence supports a hypothesis if the hypothesis explains the evidence. Of the familiar principles of qualitative confirmation abductive inference satisfies the conditions of consistency, nonuniversalizability, converse entailment, converse consequence and equivalence but it does not satisfy the conditions of entailment and special consequence. On the other hand, enumerative induction makes true the conditions of consistency, nonuniversalizability, entailment, special consequence, and equivalence but it does not make true converse entailment nor converse consequence condition. (See e.g. Smokler, 1968, for an exact formulation of these conditions.) As the reader can easily verify, apart from minor technicalities, the notion of theoretical explanation that our DEL-model makes precise (clearly) belongs to abductive inference and thus not to inductive enumeration. (One might now also argue that our DEL-model also explicates a notion of corroboration peculiar to abductive inference.) Corresponding to Smokler's notion of enumerative induction we of course also get an implicit definition of a converse relation which might be dubbed inductive systematization or inductive generalization.

The picture that emerges from this is that we have a dichotomy of theorizing both within explanation (interpretive versus phenomenological theorizing) and inductive support (abductive inference versus enumerative induction). Interpretive or theoretical explanation demonstrably goes together with abductive inference (corroboration). It can als be argued that some important forms of phenomenological theorizing go together with the above wide notion of enumerative induction and with a high probability-criterion of confirmation (cf. Niiniluoto and Tuomela, 1972; and Tuomela, 1972).

McGill University, Montreal[2]

NOTES

* I want to thank Mrs. Marja-Liisa Kakkuri-Ketonen, Mr. Seppo Miettinen, and Mr. Ilkka Niiniluoto for suggestions and criticisms concerning an earlier version of this paper.
[1] For a fuller discussion of the methodological and philosophical applications of the DEL-model of explanation see Tuomela (1972), Chapters VII and VIII.
[2] On leave of absence from the *University of Helsinki, Helsinki, Finland*.

BIBLIOGRAPHY

Ackermann, R., 1965, 'Deductive Scientific Explanation', *Philosophy of Science* **32**, 155–67.
Ackermann, R. and Stenner, A., 1966, 'A Corrected Model of Explanation', *Philosophy of Science* **33**, 168–71.
Bunge, M., 1967, *Scientific Research II*, Springer-Verlag, New York.
Bunge, M., 1968, 'The Maturation of Science', in I. Lakatos and A. Musgrave (eds.), *Problems in the Philosophy of Science*, North Holland Publishing Company, Amsterdam.
Campbell, N., 1920, *Physics: The Elements*; reprinted as: *Foundations of Science*, Dover, New York.
Campbell, N., 1921, *What is Science?*, Dover, New York.
Carnap, R., 1966, *Philosophical Foundations of Physics* (ed. by M. Gardner), Basic Books, New York.
Craig, W., 1956, 'Replacement of Auxiliary Expressions', *The Philosophical Review* **65**, 38–55.
Eberle, R., Kaplan, D., and Montague, R., 1961, 'Hempel and Oppenheim on Explanation', *Philosophy of Science* **28**, 418–28.
Hempel, C., 1969, 'Reduction: Ontological and Linguistic Facets', in S. Morgenbesser, P. Suppes, and M. White (eds.), *Philosophy, Science, and Method*, St. Martin's, New York, pp. 179–99.
Hempel, C. and Oppenheim, P., 1948, 'Studies in the Logic of Explanation', *Philosophy of Science*, 15, 135–75; reprinted in Hempel, C., 1965, *Aspects of Scientific Explanation*, The Free Press, New York, pp. 245–90.
Hintikka, J., 1970, 'Surface Information and Depth Information', in J. Hintikka and P. Suppes, (eds.), *Information and Inference*, D. Reidel Publishing Company, Dordrecht.
Hintikka, J., 1972, 'On the Different Ingredients of an Empirical Theory', forthcoming in the *Proceedings of the IVth International Congress in Logic, Methodology, and Philosophy of Science*.
Hintikka, J. and Tuomela, R., 1970, 'Towards a General Theory of Auxiliary Concepts and Definability in First-Order Theories', in J. Hintikka and P. Suppes (eds.), *Information and Inference*, D. Reidel Publishing Company, Dordrecht, pp. 92–124.
Kaplan, D. 1961, 'Explanation Revisited', *Philosophy of Science* **28**, 429–36.
Kim, J. 1963, 'On the Logical Conditions of Deductive Explanation', *Philosophy of Science* **30**, 286–91.

Lakatos, I., 1970, 'Falsificationism and the Methodology of Scientific Research Programmes', in I. Lakatos and A. Musgrave (eds.), *Criticism and the Growth of Knowledge*, Cambridge University Press, Cambridge.
Nagel, E. 1961, *The Structure of Science*, Harcourt, Brace and World, London.
Niiniluoto, I. and Tuomela, R., 1972, *Theoretical Concepts Within Inductive Systematization*, forthcoming.
Omer, I., 1970, 'On the D-N Model of Scientific Explanation', *Philosophy of Science* 37, 417–33.
Ramsey, F., 1931, *The Foundations of Mathematics and Other Logical Essays*, Littlefield, Adams and Company, Paterson, N. J.
Sellars, W., 1961, 'The Language of Theories', in H. Feigl and G. Maxwell (eds.), *Current Issues in the Philosophy of Science*, Holt, Rinehart and Winston, Inc., New York.
Smokler, H., 1968, 'Conflicting Concepts of Confirmation', *The Journal of Philosophy* 65, 300–312.
Stegmüller, W., 1969, *Wissenschaftliche Erklärung und Begründung*, Springer Verlag, Berlin, Heidelberg and New York.
Tuomela, R., 1972, *Theoretical Concepts,* Springer-Verlag, Vienna.

PART VI

METAPHYSICS

PART I

METAPHYSICS

PETER KIRSCHENMANN

CONCEPTS OF RANDOMNESS

I. INTRODUCTION

The notion of randomness has always been rather perplexing. Although it is frequently used in natural and social science, both technically and informally, it seems to have been somewhat neglected by philosophers of science ever since the discussion of the foundations of the so-called frequency theory of probability, in which it was assigned a basic role, has faded. Yet this discussion is of such significance that any attempt at clarifying the notion of randomness will have to relate to it. After a few preliminary remarks on some of the problems and puzzles of randomness, I shall, therefore, expound and discuss a concept of random distribution of a property in classes and sequences, defined in terms of relative frequencies and their limits. Because of certain shortcomings of this concept it appears advisable to turn to probabilities, in terms of which a quite different concept, viz., that of random conjunction of properties, can readily be defined as stochastic independence. This concept still has features clashing with the ordinary sense of 'randomness' which become manifest in cases where certain probabilities assume extreme values. However, when we take measures defined in information theory as measuring the degree of randomness, to which purpose they lend themselves particularly well, we find that these seemingly troublesome cases are rather harmless. A by-product of the discussion of measures of randomness is the concept of primitive randomness. The conclusion points out some further problems.

II. PRELIMINARY CONSIDERATIONS

The puzzling character of the notion of randomness can be brought out by means of some intriguing problems which have arisen in its context. Their discussion will lead to preliminary distinctions and clarifications.

M. Bunge (ed.), Exact Philosophy, 129–148. All Rights Reserved
Copyright © 1973 by D. Reidel Publishing Company, Dordrecht-Holland

II. 1. *Apparent Self-Contradiction.*

A randomizer is supposed to generate random sequences of certain elements. In repeated tossings of a fair coin, for instance, one expects to get a sequence like

(1)   10110100011...,

where '1' stands for heads, and '0' for tails. Since however, at each toss heads and tails are equally likely to occur, the following sequence is as likely as sequence (1):

(2)   1111111111...

As one would not call this sequence 'random', one would have to say that a randomizer at least sometimes generates a non-random sequence, which seems contradictory. For similar reasons G. Spencer Brown speaks of "the ultimate self-contradiction of the concept of randomness" (1957, p. 57), ultimate, because as the sequences become longer some of the features he considers become more conspicuous.

Obviously, the seeming contradiction evaporates when we distinguish between random generation and random arrangement. Any *order* exhibited by the generated sequence is compatible with the fact that the generating mechanism is a *chance* mechanism.

Besides, sequence (2) can be said to be as likely as sequence (1) only when we confine our comparison to the chances of heads and tails in each single toss. As soon as we consider other features as well, like chances for runs of a certain length, sequence (2) becomes less likely than (1).

II. 2. *The π-puzzle.*

Usually, a sequence is regarded as random if it does not possess any regular pattern, or if the constituent elements occur in it according to no rule or law. The sequence formed by the decimal expansion of the transcendental number π,

(3)   3141592653358979323846...,

not only appears to be random, but has also passed all statistical tests for randomness so far applied to it (cf. Pathria, 1962; also Stoneham, 1965). On the other hand, there are well-known mathematical formulas that can be used to compute the numerical value of π. Thus, sequence (3) can be said to be both random and governed by a rule, which seems paradoxical (cf. Venn, 1888, pp. 112f.).

The apparent paradox dissolves on realizing that again two different points of view are involved: the computational and the statistical. The formulas employed to compute the value of π do not consecutively specify each digit, nor can they be used to determine the frequencies of digits or other statistical features of the expansion of π. This is why mathematicians have to undertake what they call 'empirical' studies of this expansion. What the mathematical rules yield is rather a series of decimal expansions, infinite themselves, which converges to the value of π. On the other hand, unless we draw upon our familiarity with the expansion of π, none of the computational rules can be inferred from sequence (3). In general, any statistical structure of the generated sequence can be compatible with the fact that a generating rule yields just one determinate sequence (see, however, Section IV.2).

II. 3. *The Paradox of Random Selection.*

We customarily call the selection of an object from an aggregate of objects 'random' if each object has an equal chance of being chosen. Furthermore, we would think that the chances for an object to be so selected should not depend on how the objects are arranged in the aggregate. Yet, consider the question of the chances that a natural number picked at random will be a prime. When the natural numbers are taken in their natural order, i.e., when the aggregate in question is

(4)   1 2 3 4 5 6 7 8 9 10 11 12...,

the chance, as given by the relative frequency, of picking a prime tends to be zero. However, if we rearrange the natural numbers in the following way:

(5)   1 2 4 3 5 6 7 11 8 13 17 9 ...,

so that the aggregate consists of pairs of primes interspersed with single non-primes, then the chance of getting a prime is $\frac{2}{3}$. Accordingly, depending on the arrangement, the chances that a natural number selected at random will be a prime can vary from zero to one (cf. Russell, 1948, pp. 366f.).

The example shows that it is ambiguous to speak without further qualification of 'random selection' when the selection concerns infinite sets, be they denumerably infinite or not. An instance of the latter case is the familiar 'Bertrand's paradox' (cf., e.g., Kneale, 1949, pp. 184 ff.). The specification of the random selection involves a specification of the selection procedure together with a specification of the relevant set and its ordering (cf. Cannavo, 1966, pp. 136ff.).

II. 4. *The Puzzle of Inferences from Randomness.*

There are many successful applications of probability calculus and its laws to random or chance occurrences. In such applications statements are derived which are well corroborated. But 'random' means lawless and incalculable. How can one draw calculable conclusions from incalculabilities? (cf. Popper, 1968, p. 150)

(This question has several variants. On a subjectivist view, randomness is an expression of our ignorance. This raises the question of how we can conclude anything from what we do not know. Other variants concern ontological rather than logical relationships: How can there be chaos on a molecular level together with order on a molar level? Or, how can random mutations be at the origin of the orderly functioning and the well-adapted behavior of organisms?)

Clearly, one cannot draw conclusions from anything in random occurrences that is not amenable to logical treatment or calculation. Conclusions can be drawn, however, from hypotheses about probabilities or probability distributions, hypotheses which are assumed to hold for the occurrences in question. The statistician, on the other hand, is interested not so much in the random nature of occurences as, e.g., in their stable relative frequencies, which can be compared with hypothesized or calculated probabilities (cf. Kendall, 1941, p. 5).

II. 5. *The Problem of Collectives.*

R. von Mises took probability theory to be an emprical science, the science of 'collectives', i.e., mass phenomena, repetitive events, or potentially indefinitely long sequences of observations, which were to satisfy two conditions (see, e.g., von Mises, 1957). One of them is the principle of convergence, or the so-called limit-axiom. The other is the principle of irregularity, or the axiom of randomness, also called the 'principle of the excluded gambling system'.

The problem of collectives can be taken to include several questions, one being whether the principle of convergence makes any sense with respect to collectives. The main questions, however, have concerned the existence of collectives: Can they be shown, possibly by a constructive proof, to exist mathematically? Or, at least, can it be shown that the two conditions are compatible? Mathematicians, philosophers, and logicians have scrutinizingly dealt with these questions, without finding answers that would have complied with von Mises' original intentions or justified his approach. Since there exists an excellent review of this discussion by P. Martin-Löf (1969), I shall not go into any detail here (see also Section IV.2). With the advent of Kolmogorov's axiomatic approach to probability theory (Kolmogorov, 1933) interest in the problem of collectives has declined considerably. Now it is customary to characterize randomness in terms of probability rather than conversely (see also Section VI.3).

II. 6. *Undefinability of Randomness.*

There are conflicting views concerning the definition of a concept of randomness. R. von Mises' controversial principle of irregularity is a definition of absolute randomness in sequences. Several concepts of restricted randomness have been defined in the context of the problem of collectives. Random distribution and random conjunction of properties will be defined and discussed below. However, there are mathematicians and philosophers who have various doubts with regard to definitions of randomness. It has been said that it seems impossible to give a precise definition of what we mean by 'random', and that the sense of this term is better conveyed by examples, like tosses of a coin, combinations of genes, and death occurrences (Cramér, 1946, pp. 138ff.); or, that the concept of randomness is not extensionally definable, because it does not refer to particular elements but to the generating process for sequences (Rescher, 1961, p. 8); or, that any attempt at defining disorder in a formal way will lead to a contradiction, and that one has to appeal to something real like the urn model in order to say what disorder or random choice is (Freudenthal, 1968, p. 9).

Yet, nothing seems to preclude definitions of certain types of randomness. They may, of course, not coincide with a preconceived sense of 'randomness'. Illustrating randomness by means of examples cannot be our only resort. What is random in such examples should at least be susceptible to characterization, if not definition. What is true, though, is that a formal definition of absolute randomness, in the sense of disorder, of infinite sequences is impossible (see Section IV.2). But in this case, examples are of no avail either.

III. RANDOM DISTRIBUTION IN CLASSES AND SEQUENCES

In the following, I shall sketch a theory of random distribution drawing upon ideas elaborated by Kendall (1941), von Wright (1951, esp. ch.8), and Cannavo (1966). It possesses several features showing R. von Mises' influence. Its basic idea can easily be conveyed by an example. Take a well-specified group of people, e.g., the registered students of a college. Some of them have black hair, others do not. Now select from the group all those with blue eyes. If, in the selected subgroup, the proportion of students with black hair is the same as in the original group, the property of having black hair will be said to be randomly distributed in the original group, relative to the property of having blue eyes.

In greater detail and with some ramifications, the idea can be stated as follows. Let H be a property defining a denumerable class H'; H will be called the 'reference property', and H' the 'reference class'. If H' is finite, its members may be ordered or not. If H' is infinite, let its members be linearly ordered, in a definite way, so as to form a sequence (because of the problems discussed in Section II. 3). Let A be a property which some of the members of H' do have, i.e., $A'.H' \neq \emptyset$, where A' is the class defined by A. We shall call A the 'conjectural property', and A' the 'conjectural class'.

Def. 1. The *relative frequency* of A, given $H - F(A/H)$ for short – is defined as follows:

(a) If H' has a finite number N of members, and if n_A members of H' have property A, then $F(A/H) = n_A/N$.

(b) If H' is infinite; if n_A of the n first members of H' have property A; and if the limit $\lim_{n \to \infty} n_A/n$ exists: then $F(A/H) = \lim_{n \to \infty} n_A/n$.

Def. 2. Let S be a property. S is said to be a *selection* with respect to H' if and only if the subclass of H' consisting of the members of H' which have property S, is not empty, i.e., $H'.S' \neq \emptyset$; and in case H' is infinite, the subclass $H'.S'$ is also infinite.

If M is a material property, like the one in the introductory example, and a selection with respect to H', we shall speak of a 'material selection'. If F is a formal property and a selection with respect to H', we shall call it a 'formal selection'. Formal properties are to be attributed only to members of ordered reference classes, i.e., sequences. Formal selections refer to the places of the elements in a sequence. These places can be uniquely specified by the ordinal numbers of the elements. An example of a formal property is that of being the successor of three elements with the property A, or that of having an ordinal number exactly divisible by seven.

Since the subclass $H'.S'$ can also be considered as a reference class, the relative frequency of A, given $H.S$, i.e., $F(A/H.S)$, is defined according to Def. 1. The same holds for the relative frequency $F(S/H)$, or $F(A.S/H)$, when S, or $A.S$, is regarded as a conjectural property.

Def. 3. A is said to be *randomly distributed* in H' defined by H, *relative to S*, if and only if the relative frequency of A, given $H.S$, is equal to the relative frequency of A, given H; or symbolically

$$RD(A, S, H) \leftrightarrow F(A/H.S) = F(A/H),$$

where $F(A/H.S) = F(A/H) = f$ with $0 \leq f \leq 1$, according to Def. 1.

We shall speak of 'material randomness' or 'formal randomness' according to whether the selection is a material of a formal one. $RD(A, S, H)$ can also be expressed by saying that S is irrelevant to the distribution of A in H'. In case $F(A/H.S) > F(A/H)$, S is favorably relevant to the distribution of A in H'; and in case $F(A/H.S) < F(A/H)$, S is unfavorably relevant to the distribution of A in H'.

Def. 3 can be generalized to cases with more than one selection. Let Φ be a family of properties which are selections with respect to H'. A is said to be randomly distributed in H', relative to Φ, if and only if (S) $[S \in \Phi \rightarrow RD(A, S, H)]$. Φ can be called the 'domain of invariance' of the relative frequency of A, given H. Cannavo presupposes that the members of Φ are logically independent of each other. This condition eliminates redundancies in the domain of invariance, which then comprises only properties taken to be atomic.

Def. 4. The distribution of A in H' is said to be *absolutely random* if and only if there is no property S logically independent of A such that $F(A/H.S)$ is different from $F(A/H)$; or symbolically, with Ψ as the family of all properties which are logically independent of A, and selections with respect to H',

$$\text{ARD}(A, H) \leftrightarrow (S) [S \in \Psi \rightarrow \text{RD}(A, S, H)].$$

The requirement that S be logically independent of A, i.e., that S logically entails neither the occurrence nor the non-occurrence of A, is indispensable. For otherwise Def. 4 would be void in all but extreme cases. For example, if S logically entails A, $\text{RD}(A, S, H)$ cannot hold except in those extreme cases where all, or none, of the members of H' have property A, that is, $A' \cdot H' = H'$, or $\emptyset$. (However, S may entail A by virtue of a theory holding for the kind of things which have property H, and may have properties S and A. If this is the case, then, by Def. 4, the distribution of A in H' is just not absolutely random.)

IV. DISCUSSION OF THE NOTION OF RANDOM DISTRIBUTION

Although M. G. Kendall, who treats only of formal randomness, speaks of 'a theory of randomness', the theory sketched above is not a theory in the proper sense. It does not contain any specific primitive concept regulated by a specific axiom. The definitions given presuppose nothing beyond ordinary analysis and an algebra of classes. There are some other points which deserve more detailed consideration.

IV. 1. *Random Distribution in Finite Classes and Sequences*

A consequence of Def. 3 is that, roughly speaking, adding an element, whatever its properties, to a finite random sequence destroys the randomness. In greater detail, we have:

Thm. 1. Let K' be a reference class or sequence obtained from H' by adding or subtracting one element (with reference property H), or by replacing one member with one differing in the property A or S. Then, except in four extreme cases, if A is randomly distributed in H', relative to S, it is not so in K', and conversely.

The proof is one of simple arithmetic, using Def. 1(a). The four extreme cases are those in which $F(A/H) = F(A/K) = 0$ or 1, and $F(S/H) = F(S/K) = 0$ or 1. The reason for the great sensitivity of randomness in finite classes or sequences is, of course, the fact that the relative frequencies are proper fractions, which themselves are very sensitive to such operations on the reference class.

This sensitivity is unobjectionable as long as we are interested in exact proportions, in the composition of definite classes, or in the structure of definite linear arrangements. Adding one element to H' would, in this case, simply mean that we are considering a different class K' defined by a different property K. However, this sensitivity becomes a defect if the defined concept of random distribution is to apply to the randomness in the results of repeated trials like coin tossings. In that case we should assume that randomness once found to be present in the trial sequence is not bound to vanish with the performance of another trial.

Several ways of eliminating this defect may be proposed. (a) The practical-minded statistician will tend to weaken Def. 3 by demanding only that $F(A/H.S)$ and $F(A/H)$ have more or less the same value (Spencer Brown, 1957, p. 52). Although such an idea of approximate randomness can be given a precise formulation (cf. Kolmogorov, 1963), yielding a concept of randomness in a finite sequence different from the one defined above, it seems that it should concern the application of the concept of randomness rather than the concept itself (cf. Kendall, 1941, pp. 7f.). (b) One may take all classes and sequences to be potentially infinite. This case will presently be discussed. (c) One may think of using probabilities instead of relative frequencies for a definition of randomness. This means, however, not so much eliminating the defect mentioned as defining a different concept of randomness.

IV. 2. *Random Distribution in Infinite Sequences*

The condition, in Def. 1(b), that the limit $\lim_{n\to\infty} n_A/n$ exists, restricts the applicability of the notion of random distribution to sequences which possess such limits. For this to be the case, a sequence has to be given by a mathematical rule which specifies each element as a function of its place (ordinal number), whereas the sequences one is interested in when discussing randomness are just those which are not given by such a mathematical rule.

This is one of the shortcomings of R. von Mises' notion of collective. When requiring irregularity of his collectives, von Mises presupposed what we would call 'absolute formal randomness' of the distribution of every conjectural property. More particularly, the objection has been as follows. If there is such a rule as stated above, then it is easy to find a formal property which will yield a subsequence with relative frequencies different from those of the original sequence (cf., e.g., von Wright, 1940, p. 272). Thus, randomness in a collective can never be absolute.

As a way out of this difficulty it has been proposed that one deal with empirical sequences alone (Cannavo, 1966); but this would bring us back to finite sequences. A way of circumventing the difficulty is again to work with probabilities instead of relative frequencies and their limits. Then it is possible, as done in any textbook of probability theory, to define notions of convergence different from the concept of limit in ordinary analysis. These notions, however, do not apply to the frequentist's and empiricist's ideal of a single actual sequence, but only to whole families of sequences of a kind (see also Section VI. 3).

IV. 3. *Extreme Cases*

In the infinite sequence

(6) $\quad AOAOAOAOAOAOAO...,$

for which the limits of the relative frequencies exist, the property A is randomly distributed, relative to an infinite set of formal selections, viz., those which pick out the elements having an ordinal number that is exactly divisible by a given odd number. Furthermore, in the sequence

(7) $\quad AAAAAAAAAAAAAA...$

the distribution of A is absolutely random. This shows that the defined concept of random distribution does not coincide with the ordinary sense of 'randomness' of a sequence, which may or may not be taken as a shortcoming of the definition. What one has to conclude, however, is that the domain of invariance should not be regarded as indicating something like the degree of randomness.

Sequence (7) has already been labeled an 'extreme case', One could exclude this case by requiring $A'.H' \subset H'$, which would be parallel to the requirement $A'.H' \neq \emptyset$ made for conjectural properties in order to exclude the extreme case $A'.H' = \emptyset$.

The same holds for properties serving as selections. In every sequence, the conjectural properties are randomly distributed relative to at least one selection property, viz., the reference property, or one for which $H'.S' =$

H'. One could also exclude this extreme case by requiring $H'.S' \subset H'$, which would be parallel to the requirement $H'.S' \neq \emptyset$ made for selections.

When dealing with finite reference classes, the extreme cases can be equivalently characterized by $F(X/H) = 0$ or 1 (where 'X' stands for A or S). This is not possible with respect to infinite sequences, since $F(X/H) = 0$ (or 1) does not, in this case, imply $H'.X' = \emptyset$ (or H'), although the converse still holds.

IV. 4. *Statistical Independence*

Since the conditions $F(A/H.S) = F(A/H)$ and $F(S/H.A) = F(S/H)$ are equivalent we have:

Thm. 2. If A is randomly distributed in H', relative to S, then S is randomly distributed in H', relative to A, and conversely. Or, symbolically:

$$(A)(S)(H) [\text{RD}(A, S, H) \leftrightarrow \text{RD}(S, A, H)].$$

The two equivalent conditions above can be regarded as defining statistical independence of two properties in a reference class (although 'statistical independence' is mostly used as a synonym for 'stochastic independence' which is defined in terms of probabilities). The symmetry stated in Thm. 2 is brought out explicitly by another formulation of statistical independence, viz. $F(A.S/H) = F(A/H).F(S/H)$. In short, random distribution as here defined boils down to statistical independence.

As Thm. 2 holds also for formal selections we will have somewhat strange cases where a formal property is randomly distributed relative to another property. Besides, it has not been excluded that the conjectural property is itself a formal property. Thus, we have allowed for even stranger cases where one formal property is randomly distributed relative to another formal property. In particular, if both formal properties refer to the ordinal number of the elements of an infinite sequence, the randomness of this distribution does not at all depend on the structure of the sequence itself. Such cases are essentially of the same kind as that of sequence (6), as can be seen by taking the conjectural property, in this case, to be that of having an ordinal number which is odd. In sum, it is especially the formal properties and selections, with the concomitant requirement that the reference class be ordered, which lead to rather undesirable features of the notion of random distribution.

V. RANDOM CONJUNCTION OF PROPERTIES

In several places of the foregoing discussion it was suggested that a notion of randomness be defined in terms of probabilities instead of relative frequencies and their limits. In view of the relationship between the relative randomness of a distribution and statistical independence, I shall de-define random conjunction of properties as stochastic independence. The definition presupposes the calculus of probability, in particular that of conditional probability, in its usual axiomatic formulation; the so-called algebra of events, however, is taken to be an algebra of properties. There will be no reference to formal properties in the sense of the preceding section; we shall no longer be concerned with sequences, nor directly with classes.

Let H be again a reference property, and $A, B, \ldots$ conjectural properties which may or may not occur together with the reference property H. The non-occurrence of a property A will be designated by '$\bar{A}$'. Let the conditional probabilities $P(A/H)$ – i.e., the probability of A, given H –, $P(A/H.B)$, $P(A.B/H)$, etc. be defined as usual.

Def. 5. A and B are said to be *randomly conjoined* on H if and only if A and B are stochastically independent on H; or symbolically:

$$RC(A, B, H) \leftrightarrow P(A.B/H) = P(A/H).P(B/H).$$

As can be seen from the relationships between probabilities, the following theorems hold:

Thm. 3. $(A)(B)(H) [RC(A, B, H) \leftrightarrow RC(B, A, H)]$,
Thm. 4. $(A)(B)(H) [RC(A, B, H) \leftrightarrow RC(\bar{A}, B, H) \leftrightarrow$
    $\leftrightarrow RC(A, \bar{B}, H) \leftrightarrow RC(\bar{A}, \bar{B}, H)]$,
Thm. 5. $(A)(B)(H) [P(B/H) \neq 0 \rightarrow (RC(A, B, H) \leftrightarrow$
    $\leftrightarrow P(A/H.B) = P(A/H))]$.

In analogy to Def. 4 we can say that A is absolutely random on H if and only if A is randomly conjoined with any other property logically independent of A. The relation $RC(A, B, H)$ can also be expressed by saying that A and B are, on H, irrelevant to each other. In case $P(A.B/H) > P(A/H).P(B/H)$, they are said to be favorably relevant to each other; in case $P(A.B/H) < P(A/H).P(B/H)$, they are unfavorably relevant to each other.

VI. REMARKS ON THE NOTION OF RANDOM CONJUNCTION

Corresponding to the definition above, von Wright has defined concepts of relative and absolute chance. He calls $P(A/H)$ the chance, relative to G, that a positive instance of H will be a positive instance of A if $P(A/H.G) = P(A/H)$ (von Wright, 1951, pp. 226f.). There are others who regard stochastic independence as characterizing, or being a criterion of, randomness or chance (cf., e.g., Keynes, 1921, p. 287; Bohm and Schützer, 1955, p. 1039).

VI. 1. *Relation Between Random Distribution and Random Conjunction*

From the discussion in Section II. 1 it is clear that the concept of random distribution concerns randomness in the sense of disorder, whereas that of random conjunction concerns randomness in the sense of chance. Still von Wright claims that "(relative and absolute) random distribution implies (relative and absolute) chance.... An assertion of random distribution is, however, somewhat stronger than an assertion of chance.... 'normally', *i.e.*, this extreme case [the case $P(\bar{G}/H) = 1$] being excluded, (relative and absolute) chance implies random distribution" (1951, pp. 229f.). These claims are based on the unjustified belief that the so-called frequency interpretation of probability provides the means for unambiguously inferring probabilities from relative frequencies and their limits, and, with the exception of extreme cases, also vice versa. The extreme case excluded by von Wright concerns the equivalence of two conditions for random conjunction, as expressed in Thm. 5, rather than the transition from relative frequencies to probabilities, or from chance to random distribution. The actual relation between random distribution and random conjunction is rather methodological than logical. A random distribution may lead us to suppose random conjunction; and a hypothesis of random conjunction may be tested by checking distributions of properties for randomness (cf. also Bunge, 1956).

VI. 2. *Series of Reference Properties*

In analogy to Def. 5, concepts of random conjunction can be defined in other cases of stochastic independence as well, cases which are dealt with in any textbook of probability theory. A simple example is the case of what are called 'repeated trials under identical conditions' (cf. Feller, 1957, pp. 118f.). These form a series of repetitions of the same reference property H, together with the same set of mutually exclusive conjectural properties A_j, and the same conditional probabilities $P\ (A_j/H)$. Consider, for the sake of brevity, a series of only two instances of this kind; the reference property of the series as a whole will then be the ordered pair (H, H), associated with the conjectural properties (A_j, A_k), and the probabilities $P\{(A_j, A_k)//(H, H)\}$. For this case, we have:

Def. 6. A_j in the first instance and A_k in the second instance are said to be randomly conjoined on (H, H) if and only if $P\ \{(A_j, A_k)/(H, H)\} = P\ (A_j/H) \cdot P\ (A_k/H)$.

One speaks of 'independent trials' if this condition is satisfied by all pairs (A_j, A_k). Thus, the possible outcomes of independent trials are randomly conjoined. As mentioned in the preceding section, this does not mean that the sequence of actual outcomes will exhibit random distribution of conjectural properties.

VI. 3. *Collectives Reconsidered*

With respect to infinite series of independent trials (i.e., not with respect to their outcomes alone), two theorems can be proven which are reminiscent of R. von Mises' conditions for collectives. The one is the familiar strong law of large numbers, which says, roughly speaking, that the possible relative frequencies of conjectural properties in unlimited series of independent trials under identical conditions converge almost always (or, with probability one) to the probabilities of conjectural properties in one trial. The possible relative frequencies are themselves defined in terms of the latter probabilities. The theorem does not concern the behavior of relative frequencies in actual sequences. Furthermore, the notion of convergence almost always is quite different from the concept of limit in ordinary analysis.

The other theorem is a modified version of the principle of the impossibility of a successful gambling system (cf. Feller, 1957, pp. 185f.). The systems considered are essentially such that whether or not a bet is made at a certain trial is stochastically independent of that trial. For this reason, probabilities have to be assigned to the occurrences of bets, which shows that gambling systems of this kind are different from the formal selections considered by von Mises. The theorem itself states that under any such system the successive trials betted at form a series of trials with unchanged probability for success. For all subseries of this kind the law of large numbers holds as well. W. Feller concludes that "the two theorems together describe the fundamental properties of randomness which are inherent in the intuitive notion of probability and whose importance was stressed with special emphasis by von Mises" (1957, p. 191). Aside from the fact that the two theorems are, as I have indicated, fundamentally different from von Mises' principles, the notion of random conjunction is evidently more basic than what is expressed by the two theorems, since both of them are derived under the assumption of stochastic independence.

VI. 4. *Extreme Cases*

The concept of random conjunction also has some apparently undesirable features, analogous to those of the concept of random distribution, which are again linked to extreme cases.

Thm.6. If $P(A/H)=0$ or 1, then A is randomly conjoined with any other conjectural property on H (i.e., A is absolutely random on H).

The feeling that this consequence is quite inappropriate might be alleviated by the observation that there is no contradiction involved in saying that A is absolutely random on H, but nevertheless determined by another property, which may be H itself. We shall presently see that there is a more stringent way of rendering the absolute randomness of extreme cases harmless.

Obviously, Thm. 6 implies that every conjectural property is randomly conjoined on H with any property entailed by H, hence with H itself. The latter can be taken to express the common idea that an assumption of some kind of randomness (which will be called 'primitive randomness' below) is implied whenever we speak of probabilities, i.e. not only when we speak of stochastic independence.

VII. MEASURES AND SOME FURTHER NOTIONS OF RANDOMNESS

Some of the measures defined in information theory can be taken as measures of randomness, as concerns both conjunctions of properties on a single reference property and conjunctions of properties in series of reference properties. Only the former case will be considered here.

Moreover, their mathematical properties and interrelations suggest that they be regarded as measuring degrees of randomness, not only in cases of random conjunction of properties, but also in cases where conjectural properties are not randomly conjoined, and even with respect to the probability distributions of single conjectural properties.

I shall keep with the stipulations preceding Def.5. Furthermore, '$\mathscr{A}$' will designate the probability distribution $p_X =_{df} P(X/H)$ associated with the variable X, where $X \in \{A, \bar{A}\}$. '$\mathscr{A}^*$' will designate the special case, in which $p_A = p_{\bar{A}} = \frac{1}{2}$. '$\mathscr{B}$' will stand for the probability distribution $p_Y =_{df} P(Y/H)$ associated with $y \in \{B, \bar{B}\}$. Other abbreviations are $p_{X/Y} =_{df} P(X/H.Y)$, and $p_{XY} =_{df} P(X.Y/H)$, with $p_{X/Y} \cdot p_Y = p_{XY}$. The following notions are defined in information theory (cf. Rényi, 1966, pp. 440ff.).

Defs. 7.
(a) (amount of) information $I(\mathscr{A}) =_{df} \sum_{X \in \{A, \bar{A}\}} p_X \cdot \log_2(1/p_X)$
(b) joint information $I((\mathscr{A}, \mathscr{B})) =_{df} \sum_{X \in \{A, \bar{A}\}} \sum_{Y \in \{B, \bar{B}\}} p_{XY} \cdot \log_2(1/p_{XY})$
(c) conditional information $I(\mathscr{A}/\mathscr{B}) =_{df} I((\mathscr{A}, \mathscr{B})) - I(\mathscr{B})$
(d) relative information $I(\mathscr{A}, \mathscr{B}) =_{df} I(\mathscr{A}) + I(\mathscr{B}) - I((\mathscr{A}, \mathscr{B}))$.

All of these functions are non-negative. Def. 7(a) entails that $I(\mathscr{A}^*) = 1$. The following theorems, proved in information theory (cf. Rényi, 1966, pp. 447ff.), are of particular interest in the present context.

Thms. 7. For every $\mathscr{A}$ and $\mathscr{B}$:

(i) $\quad I(\mathscr{A}) \leq I(\mathscr{A}^*)$
(ii.a) $\quad I((\mathscr{A}, \mathscr{B})) \leq I(\mathscr{A}) + I(\mathscr{B})$
(ii.b) $\quad I(\mathscr{A}/\mathscr{B}) \leq I(\mathscr{A})$
(ii.c) $\quad 0 \leq I(\mathscr{A}, \mathscr{B})$
(iii) $\quad I(\mathscr{A}, \mathscr{B}) \leq \min(I(\mathscr{A}), I(\mathscr{B}))$.

VII.1. *Random Conjunction of Properties Redefined.*

In Thms. 7(ii), the equality signs hold precisely in the special case when $\mathscr{A}$ and $\mathscr{B}$ are stochastically independent on H, i.e., by virtue of Thm. 4, when A and B are randomly conjoined on H. Hence we can replace Def.5 by any one of the following definitions.

Defs. 5'. (a) $RC(A, B, H) \leftrightarrow I((\mathscr{A}, \mathscr{B})) = I(\mathscr{A}) + I(\mathscr{B})$
(b) $RC(A, B, H) \leftrightarrow I(\mathscr{A}/\mathscr{B}) = I(\mathscr{A})$
(c) $RC(A, B, H) \leftrightarrow I(\mathscr{A}, \mathscr{B}) = 0$.

VII.2. *Conditional and Joint Randomness.*

The range of the conditional information $I(\mathscr{A}/\mathscr{B})$ is given by $0 \leqslant I(\mathscr{A}/\mathscr{B}) \leqslant I(\mathscr{A})$. It takes on its maximum in case $\mathscr{A}$ and $\mathscr{B}$ are stochastically independent, i.e. A and B randomly conjoined. It takes on its minimum when $p_{A/B}=0$ or 1, and $p_{A/\bar{B}}=0$ or 1. (Note that two of these four cases, viz. $p_{A/B}=p_{A/\bar{B}}=0$ or 1, are cases of stochastic independence). $I(\mathscr{A}/\mathscr{B})$ can be said to measure the degree of conditional randomness, i.e., the degree of randomness in $\mathscr{A}$, given $\mathscr{B}$. This means, firstly, that the degree of conditional randomness involved in random conjunctions of properties is measured by $I(\mathscr{A}/\mathscr{B})_{RC}=I(\mathscr{A})$. Secondly, this implies that we shall also speak of conditional randomness, as measured by $I(\mathscr{A}/\mathscr{B})$, in cases where $\mathscr{A}$ and $\mathscr{B}$ are not stochastically independent. In these cases, the conditional randomness can be said to be only partial, whereas in cases of stochastic independence it may be said to be complete. It should be clear that the distinction between complete and partial conditional randomness is not one in terms of degrees of conditional randomness. The degree of complete conditional randomness, i.e. $I(\mathscr{A}/\mathscr{B})_{RC}$, can vary itself. Its range is the same as that of $I(\mathscr{A})$, which is given by $0 \leqslant I(\mathscr{A}) \leqslant I(\mathscr{A}^*)$. It is smallest in case $p_{A/B}=0$ or 1, and greatest for $p_{A/B}=\frac{1}{2}$ (where, as in all cases of stochastic independence, $p_{A/B}=p_{A/\bar{B}}=p_A$). In sum, the degree of conditional randomness itself is greatest in case of stochastic independence plus equiprobability.

The joint information $I((\mathscr{A}, \mathscr{B}))$ has properties analogous to those of the conditional information, and can be said to measure the degree of joint randomness. Its range is given by $0 \leqslant I((\mathscr{A}, \mathscr{B})) \leqslant I(\mathscr{A})+I(\mathscr{B})$. As it is a symmetrical function with respect to $\mathscr{A}$ and $\mathscr{B}$, the degree of complete joint randomness, i.e. $I((\mathscr{A}, \mathscr{B}))_{RC}=I(\mathscr{A})+I(\mathscr{B})$, can be considered the most appropriate measure of the randomness involved in random conjunctions of two properties.

If one would not allow for partial randomness, as measured by $I(\mathscr{A}/\mathscr{B})$ or $I((\mathscr{A}, \mathscr{B}))$, then so-called random processes would not be random at all. Since the stages of processes depend on one another, they are, except in limiting cases, not stochastically independent, and hence not randomly conjoined. Nevertheless, processes can exhibit degrees of partial conditional, or joint, randomness.

VII.3. *Stochastic Dependence.*

The relative information $I(\mathscr{A}, \mathscr{B})$, whose range is given by $\min(I(\mathscr{A}), I(\mathscr{B})) \geqslant I(\mathscr{A}, \mathscr{B}) \geqslant 0$, has a peculiar status, because it is minimum in case of stochastic independence, and increases when conditional and joint information decrease. It is a measure of the stochastic dependence of $\mathscr{A}$ and $\mathscr{B}$, or, as it were, of the degree of non-randomness of $\mathscr{A}$ and $\mathscr{B}$ relative to one another. It is maximum when one of the properties A and B is determined by the other. In particular, $I(\mathscr{A}, \mathscr{B}) = I(\mathscr{A})$ if $p_{A/B} = 0$ and $p_{A/\bar{B}} = 1$, or $p_{A/B} = 1$ and $p_{A/\bar{B}} = 0$.

VII.4. *Primitive Randomness.*

The fact that the amount of information occurs in such relationships with the other measures of information as were stated and used in the foregoing suggests that it be also taken as measuring the degree of some kind of randomness. It certainly does so in cases of stochastic independence, where, as we have found, the degree of conditional randomness is measured by $I(\mathscr{A}/\mathscr{B})_{RC} = I(\mathscr{A})$. But as it has been deemed plausible to speak of randomness also in cases other than those of stochastic independence, we may regard $I(\mathscr{A})$, in general, as measuring the degree of primitive randomness in $\mathscr{A}$. Its range is given by $0 \leqslant I(\mathscr{A}) \leqslant I(\mathscr{A}^*)$. It is minimum for $p_A = 0$ or 1; and it is maximum when the alternatives A and $\bar{A}$ are equiprobable. The latter matches the common view that the occurrence of a property is most random when its occurrence and its non-occurrence are equally likely.

VII.5. *Extreme Cases Reexamined.*

The cases where the degree of primitive randomness is minimum are precisely the extreme cases discussed in Section VI.4. In those cases, the property A was seen to be randomly conjoined with any other property B on H, or absolutely random on H, which seemed to clash with the ordinary understanding of randomness. Now this seeming oddity disappears insofar as, in these cases, the degree of primitive randomness $I(\mathscr{A})$ as well as the degree of conditional randomness $I(\mathscr{A}/\mathscr{B})$ are zero. In short, although the randomness is absolute, it is of vanishing degree. It is only the joint randomness whose degree will, in general, not be zero. However, it reduces, in these cases, to nothing but the primitive randomness in $\mathscr{B}$.

VIII. CONCLUSION

I have discussed a concept of random distribution of properties in classes and defined a concept of random conjunction of properties; I have also discussed measures of various kinds of randomness. In concluding, I shall only mention some further problems which await treatment. Both the concept of random conjunction and the measures of randomness rest upon the notion of probability, which was not explicitly dealt with in this paper. Since, however, assumptions of randomness are frequently brought forward as justifications for working with probabilities, the relationship of randomness and probability should be examined in detail. Another task is to show which concepts of randomness are relevant in the sciences. It seems that the concept of random distribution of properties in classes, though interesting in itself and of relevance to statistics, has no use in theories of the empirical sciences. However, the assumption of randomness in the sense of stochastic independence, very often combined with the notion of randomness in the sense of equiprobability, or maximum primitive randomness, undoubtedly plays a role in various fields of science. I do not know of any case where the measures of randomness are directly employed; the idea of stochastic dependence and that of probabilities other than equiprobabilities are, of course, widely used. Finally, some of the assumptions of random conjunction of properties made in the sciences seem to be justifiable in terms of an actual independence of the properties. This raises the question as to whether, and to what extent, the notion of an actual independence is fundamental to concepts of randomness.

ACKNOWLEDGEMENTS

This paper was written while I was on a leave of absence from Wayne State University and had a grant from the Canada Council. I should like to thank Prof. M. Bunge, director of the Foundations and Philosophy of Science Unit at McGill University, for many fruitful discussions.

McGill University, Montreal[1]

NOTE

[1] Presently at Monteith College, Wayne State University, Detroit, U.S.A.

BIBLIOGRAPHY

Bohm, D. and Schützer, W., 1955, 'The General Statistical Problem in Physics and the Theory of Probability', *Nuovo Cimento Suppl.* **2**, ser. 10, 2nd sem., 1004–47.
Bunge, M., 1956, 'A Critique of the Frequentist Theory of Probability', *Congresso Internacional de Filosofia*, São Paulo, Vol. 3, 787–92.
Cannavo, S., 1966, 'Extensionality and Randomness in Probability Sequences',*Philosophy of Science* **33**, 134–46.
Cramér, H., 1946, *Mathematical Methods of Statistics*, Princeton University Press, Princeton.
Feller, W., 1957, *An Introduction to Probability Theory and Its Applications*, **1**, John Wiley, New York, London.
Freudenthal, H., 1968, 'Realistic Models in Probability', In I. Lakatos (ed.), *The Problem of Inductive Logic*, North Holland, Amsterdam, pp. 1–14.
Kendall, M. G., 1941, 'A Theory of Randomness', *Biometrika* **32**, 1–15.
Keynes, J. M., 1921, *A Treatise on Probability*, MacMillan, London.
Kneale, W., 1949, *Probability and Induction*, Clarendon, Oxford.
Kolmogoroff, A. N., 1933, *Grundbegriffe der Wahrscheinlichkeitsrechnung*, Springer, Berlin.
Kolmogorov, A. N., 1963, 'On Tables of Random Numbers', *Sankhya* **25**, 369–76.
Martin-Löf, P., 1969, 'The Literature on von Mises Kollectivs Revisited', *Theoria* **35**, 12–37.
von Mises, R., 1957, *Probability, Statistics and Truth*, MacMillan, New York.
Pathria, R. K., 1962, 'A Statistical Study of Randomness Among the First 10000 Digits of π', *Mathematics of Computation* **16**, 188–97.
Popper, K. R., 1968, *The Logic of Scientific Discovery*, Harper Torchbook, New York.
Rényi, A., 1966, *Wahrscheinlichkeitsrechnung*, Deutscher Verlag der Wissenschaften, Berlin.
Rescher, N., 1961, 'The Concept of Randomness', *Theoria* **27**, 1–11.
Russell, B., 1948, *Human Knowledge*, Simon and Schuster, New York.
Spencer Brown, G., 1957, *Probability and Scientific Inference*, Longmans, London, New York, Toronto.
Stoneham, R. C., 1965, 'A Study of 60000 Digits of the Transcendental 'e'', *American Mathematical Monthly* **72**, 483–500.
Venn, J., 1888, *The Logic of Chance*, Chelsea, New York.
von Wright, G. H., 1940, 'On Probability', *Mind* **49**, 265–83.
von Wright, G. H., 1951, *A Treatise on Induction and Probability*, Routledge and Kegan Paul, London.

PART VII

ETHICS

BAS C. VAN FRAASSEN

THE LOGIC OF CONDITIONAL OBLIGATION

Various paradoxes in deontic logic have led to the introduction of concepts of conditional obligation. The aim of this paper is to develop a semantic theory of conditional obligation, a complete logical system pertaining thereto, and a translation into modal logic analogous to that provided by Anderson for normal deontic logics.

I. ABSOLUTE OBLIGATIONS

Accepting the general obligation to bring about whatever ought to be the case, and the thesis that what ought to be the case is exactly what is the case in any ideal situation, deontic logicians have devised a minimal deontic logic variously called D, DL (Åqvist)[1], or DM (Fitch)[2]. In axiomatic form, it has (with O read as "it ought to be the case that").

A1 Axiom schemata for propositional calculus
A2 $\vdash O(A) \supset {\sim} O({\sim} A)$
A3 $\vdash O(A \supset B) \supset . O(A) \supset O(B)$
R1 If $\vdash A$ and $\vdash A \supset B$ then $\vdash B$
R2 If $\vdash A$ then $\vdash O(A)$

It is not surprising that this simple system allows the formalization of only a narrow fragment of discourse concering duties and obligations.

Of the various problems not handled, perhaps the most important is that of contrary-to-duty imperatives.[3] Granted that one ought not to steal, the obligation to make restitution if one does steal is one that arises when another obligation has been violated. But within deontic logic constructed along the lines indicated above, statements of such obligations are not adequately formulable.

For $O({\sim} A)$ implies $O(A \supset B)$ no matter what B is, and $A \supset O(B)$ contradicts the conjunction of $O({\sim} A)$, A, and $O({\sim} B)$.

M. Bunge (ed.), Exact Philosophy, 151–172. *All Rights Reserved*
Copyright © 1973 by D. Reidel Publishing Company, Dordrecht-Holland

II. CONDITIONAL OBLIGATIONS

In answer to problems of the kind mentioned above, von Wright proposed that the concept of an obligation to do something *given* certain conditions is not definable in terms of a concept of obligation *simpliciter*.[4] He proposed as minimal criterion for a logic of conditional obligations that absolute obligations should be definable as obligations conditional on tautologous conditions. That is, if we introduce the dyadic operator O, regarding $O(A/B)$ as stating that under conditions satisfying B, A ought to be satisfied, the monadic operator of system D should be definable by the equivalence $O(A) = O(A/B \supset B)$.

In the same article, von Wright suggested two axiom schemes for conditional obligations. He stated these in terms of what is permitted rather than what is obligatory. If we follow the usual course of defining "it is permitted that..." as "it is not obligatory that not..." these axiom schemes are

(1) $\vdash O(A/C) \supset \sim O(\sim A/C)$
(2) $\vdash O(A \vee B/C) \equiv O(A/C) \,\&\, O(B/C \,\&\, \sim A)$

In restating these axioms I have assumed not only the usual definition of permission but also the rule that if A and B are provably equivalent, then so are $O(A/C)$ and $O(B/C)$. From the article it is actually not clear what von Wright assumed concerning the logical apparatus beyond his axiom schemes.

If von Wright assumed the obvious and minimal generalization of A3 and R2, namely

R2   If $\vdash A \supset B$ then $\vdash O(A/C) \supset O(B/C)$

then we can deduce the following theorem

T $\vdash O(A/C) \supset O(A/C \,\&\, \sim B)$

(For $A \vdash B \vee A$, so $O(A/C) \vdash O(B \vee A/C)$. But by (2.) $O(B \vee A/C) \vdash O(A/C \,\&\, \sim B)$.). Now this consequence is made unacceptable by the problem of obligations overridden by new circumstances.

This problem was raised in an example that Powers constructed *à propos* a system due to Åqvist.[5] I quote a short formulation: Powers "gives the example of John Doe and Suzy Mae who violated a primary obligation. Due to the violation of this primary obligation a secondary obligation takes over, that of John marrying Suzy Mae. This is not all because John has violated another primary obligation by shooting Suzy Mae...so John cannot marry Suzy. Hence he does not have a secondary obligation to marry Suzy".[6] The main point is that, although given conditions C, it ought to be the case that A, conditions C & B may make it quite impossible for A to be the case.

In a later work,[7] von Wright suggests as a 'natural' axiom scheme for conditional obligation (in effect):

(3) $\vdash O(A \,\&\, B/C \vee D) \supset O(A/C) \,\&\, O(B/C) \,\&$
       $O(A/D) \,\&\, O(B/D)$

But assuming what von Wright calls a "rule of extensionality", that sentences provably equivalent in the propositional calculus may be substituted for each other everywhere, T can be deduced again. (For (3.) yields $\vdash O(A/(C \,\&\, B) \vee (C \,\&\, {\sim} B)) \supset O(A/C \,\&\, {\sim}B)$. and $(C \,\&\, B) \vee (C \,\&\, {\sim} B)$ is tautologously equivalent to C.). Thus it seems that von Wright's proposals for the logic of conditional obligation run afoul of the John and Suzy paradox, if R2' be accepted. (And the John and Suzy example is but one of a large family: questions should be answered truthfully, but not if a truthful answer will help to make a crime succeed; the 'everything else being equal' clause that tacitly accompanies statements of conditional obligation cannot be removed.) But R2' is not easily given up.[8]

To argue for the retention of R2', I shall outline an interpretation of $O(A/B)$ suggested by, but rather wider than, the interpretations considered informally by Powers. In the interpretation of statements of absolute obligation we use the following picture: a certain set of possible worlds is specified as ideal, and $O(A)$ is true in the actual world exactly if A is true in all ideal worlds. We can regard $O(A)$ as playing a role in the evaluation of our world, ("Stealing ought not to happen, but it does; that is bad.") or in decision making ("our decisions realize various possible states for the world tomorrow; aim to produce an ideal state."). Staying with the second of these, we can liberalize the picture: with respect to tomorrow our choice is not merely to actualize an ideal state or a non-ideal state, but to actualize a more or less ideal state. That is, each possible outcome of our decisions today has a certain value. Now suppose that due to facts beyond our control or prior decisions, C will be the case tomorrow. Then we can only aim for higher values within the set of states that satisfy C. And so $O(A/C)$ presumably means something like "Given C, to maximize value requires that A".

I said "something like", for we have to consider various possibilities. Let us designate the set of attainable states satisfying a given sentence S as $H(S)$. Then if there is a maximum among the values of states in $H(C)$, $O(A/C)$ is true just if that maximum lies in $H(A \& C)$. If there is no maximum, then we should compare $H(A \& C)$ with $H(\sim A \& C)$. If for every value in $H(A \& C)$ there is one at least as high in $H(\sim A \& C)$, $O(A/C)$ is *not* true. More concisely, $O(A/C)$ is true if some state or world in $H(A \& C)$ has a value higher than any to be found in $H(\sim A \& C)$. To put it still another way: $O(A/B)$ is true exactly if opting for $\sim A \& B$ precludes the attainment of some value which it is possible to attain if one opts for $A \& B$. But does this not ignore the problem of likelihood? Is gambling the most moral of pursuits if breaking the bank makes possible unrivalled philanthropy? I don't mean that of course. In assigning values to possible outcomes relative likelihood must be taken into account; this is an old theme of decision theory. And indeed, an old theme of morals: the gambler who loses his wages is culpable vis-à-vis his dependents even if all his winnings would have been spent to their benefit.

Before examining the John and Suzy paradox anew, let us scrutinize the relation between 'ought' and 'better', as Åqvist did in connection with absolute obligations.[9] The set of outcomes that satisfy A is *better* than the set that satisfy B if some element of the former has a value higher than any fount in the latter; symbolically, $B(A/B)$. But then $O(A/B)$ is, by our account, exactly equivalent to $B(A \& B/\sim A \& B)$. Now the von Wright theorem that runs afoul of this paradox is $O(A/B) \supset O(A/B \& C)$. This is then exactly equivalent to $B(A \& B/\sim A \& B) \supset B(A \& B \& C/\sim A \& B \& C)$. But that is easily refuted. For example, if $H(A \& B \& C)$ is empty, it cannot be better than anything. Or, more mundanely, if somewhere in $H(A \& B)$ we see a high value, that value might nevertheless not lie in $H(A \& B \& C)$. (There is an easy procedure for checking this: draw a Venn diagram and write number variables in the compartments.)[10]

III. CRITERIA FOR A LOGIC OF CONDITIONAL OBLIGATIONS

The criterion proposed by von Wright is that in the logic of conditional obligations, the monadic operator O defined by $O(A) \equiv O(A/B \supset B)$ should satisfy system D.[11]

I wish to strengthen this criterion: we should be able to demonstrate that if a sentence B is added as an axiom, then in the extended system the monadic operator O^B defined by $O^B(A) \equiv O(A/B)$ should satisfy system D. After all, in that extended system, B has the status of $B \supset B$.

As a further criterion, I propose that if something is a necessary condition of discharging an obligation then it is itself an obligation, given the same conditions. This is clearly the rule discussed in the preceding section.

Together these criteria leave much undetermined, since they say nothing about how different conditions are related. Before going on to that problem, let us state the system CD – as we shall call the logic of conditional obligations – to the extent that it can now be determined.

AC 1   Axiom schemata for propositional logic.
AC 2 $\vdash O(A/C) \supset \sim O(\sim A/C)$
AC 3 $\vdash O(A \supset B/C) \supset . O(A/C) \supset O(B/C)$
RC 1   if $\vdash A$ and $\vdash A \supset B$, then $\vdash B$
RC 2   if $\vdash A \supset B$ then $\vdash O(A/C) \supset O(B/C)$
RC 3   if $\vdash A$ then $\vdash O(A/A)$

Obviously, without rule RC 3 the assumption that $\vdash B$ leads only to $\vdash O(B/B) \supset O(A \supset A/B)$; and to satisfy our criteria, we must be able to prove $O^B(A \supset A)$ on the assumption that $\vdash B$. All of AC 1–3 and RC 1–2 are directly demanded by our criteria, and RC 3 is a minimal addition to guarantee that the criteria are entirely satisfied.

There is a rudimentary semantic criterion that yields another axiom and rule. The intuitive meaning of "given A" is such that, if a sentence ends with it, then any possibility that does not satisfy A is irrelevant to the evaluation of that sentence. Thus, the evaluation of $O(A/B)$ cannot depend on $H(A)$ as such but at most on $H(A \& B)$. Succinctly: there must be a relation R such that $O(A/B)$ is true exactly if $H(A \& B)$ bears R to $H(B)$. This criterion seems to me to be largely independent of our special interpretation, and does not require acceptance of the axiological slant of our current approach. But it entails directly the necessity for the following additions to the logical system.

AC 4 $\vdash O(B/A) \supset O(B \& A/A)$
RC 4   If $\vdash C \equiv D$ then $\vdash O(A/C) \equiv O(A/D)$

We require, finally, a set of axioms that do reflect that axiological slant. Recall the relation *better*: we said that $O(A/B)$ is true exactly if $B(A \& B/\sim\sim A \& B)$ is. With very little ingenuity (which I shall take pains to display when I formalize the semantic account) it can be shown that $B(A/B)$ is in turn equivalent to $O(\sim B/A \vee B)$. So we add a definition to this effect, and then axioms which will ensure that 'better' has all its intuitively rightful properties (such as transitivity):

Definition '$B(A/B)$' for '$O(\sim B/A \vee B)$'
AC 5 $\vdash B(A/B) \supset [B(B/C) \supset B(A/C)]$
AC 6 $\vdash \sim B(A/B) \supset [B(A/C) \supset B(B/C)]$
AC 7 $\vdash \sim B(A/B) \supset [B(C/B) \supset B(C/A)]$

For future reference, we list some theorems.

T 1 $\vdash O(A \vee \sim B/B) \equiv O(A/B)$

For $O(A/B) \supset O(A \vee \sim B/B)$ by RC 2; $O(A \vee \sim B/B)$ implies $O[(A \vee \sim B) \& B/B]$ by AC 4 and hence $O(A/B)$ by RC 2.

T 2 $\vdash O(A/C) \& O(B/C). \supset O(A \& B/C)$

For $O(B/C) \supset O(A \supset A \& B/C)$ by RC 2, which together with $O(A/C)$ implies $O(A \& B/C)$ by AC 3.

T 3 $\vdash \sim O(\sim A/A)$

For suppose $O(\sim A/A)$; then $O(A \& \sim A/A)$ by AC 4. But then $O(A/A)$ by RC 2 and hence $\sim O(\sim A/A)$ by AC 2.

T 4 $\vdash O(A/B) \equiv B(A \& B/\sim A \& B)$

For $B(A \& B/\sim A \& B) \equiv O(A \vee \sim B/B) \equiv O(A/B)$ by T 1.

IV. PRACTICAL ACTION AND THE PARADOXES

Suppose that one considers what is to be done, with an eye on the moral values of the possible outcome of one's actions. Then if one knows that the actual outcome must satisfy C, and that $O(B/C)$ is true, ought one to follow a course of action leading to an outcome that satisfies B? The answer is "no, not necessarily"; for example one may know as well that courses of action leading to outcomes satisfying B are not possible. This is clearly the lesson of the John and Suzy paradox.

We have a problem here analogous to the problem of detachment for conditional probabilities. And I propose that the former be solved analogous to the solution proposed by Carnap for the latter: by separating the principles for the application of the calculus from the principles of the calculus, and imposing a 'total evidence' requirement. Thus suppose that what we know will be the case tomorrow regardless of our actions can be summed up exactly in statement A. Then, if $O(B/A)$ is true, it is to be accepted that we ought to follow a course of action that leads to an outcome satisfying B – or at least that we ought to try. Our calculus makes this maxim consistent, since $O(B/A)$ & $O(\sim B/A)$ cannot be deduced. The maxim is not helpful to one whose knowledge cannot be finitely axiomatized in his own language, but such a person would in any case be well advised to switch to a language with greater resources.

Thus we distinguish between practical judgments (injunctions or mandates) and theoretical judgments of obligation. The question what is to be done may be answered by the practical judgment that X ought to be done. But this practical judgment does not state an unconditional obligation; it is warranted or justified by a theoretical judgment that it ought to be that X be done, given the known conditions. This theoretical judgment alone can be expressed in our language. And every theoretical judgment carries a *ceteris paribus* rider. "Thou shalt not kill" either states an unconditional obligation or is a practical judgment warranted by one. The statement of unconditional obligation can be expressed by "It is (morally) better not to kill than it is to kill (*ceteris paribus*)". This leaves it open that under certain special conditions (defense of one's virtue, say) it is morally justified to kill. Unconditional statements of obligation, like Aristotle's universal statements, are normally subject to exceptions.

As Åqvist already pointed out, the Good Samaritan paradox is a special case of the problem of contrary-to-duty imperatives.[12] Given that a man has been robbed, we are obligated to help a man who has been robbed; but simpliciter, it is better that there be no man who has been robbed. But I would like to consider briefly the use of this paradox by Castañeda to critize a certain principle (that looks somewhat like our RC 2).[13] Suppose that Robert is the man whom Benjamin robs, and that Arthur is obligated to bandage Robert. Then is ought to be that Arthur bandage the man Benjamin robs. But that Arthur bandages the man Benjamin robs implies that Benjamin robs some one. Hence it ought to be that Benjamin rob someone. This inference must clearly be rejected.

Prima facie, what is at fault is that D has both A 3 and R 2, which together yield that if $\vdash A \supset B$ then $\vdash O(A) \supset O(B)$. And if this is indeed the exact location of the fallacy, then our RC 2 would also be impugned. But I think that rather, Castañeda's example shows again that certain inference patterns cannot be adequately represented in D (to which, and like calculi, his criticism was addressed). Let us formalize the argument, using the obvious symbols:

(1) $r = (\imath x)(Rbx)$
(2) $O(Bar)$
(3) $O(Ba(\imath x)(Rbx))$
(4) $\vdash Ba(\imath x)(Rbx) \supset (Ex)(Rbx)$
(5) $O((Ex)Rbx)$.

The move from (1) and (2) to (3) by substitutivity of identity is not warranted in D, in which (1), Bar, and $Ba(\imath x)(Rbx)$ are atomic statements. But the following principle may be assumed,

(0) $\vdash r = (\imath x)(Rbx) \ \& \ Bar. \supset Ba(\imath x)(Rbx)$

Hence the principle formulable in D which is rejected is that which leads from (0), (1), and (2) to (3), namely

(6)   if $\vdash A \ \& \ B \supset C$ then $\vdash A \ \& \ O(B). \supset O(C)$

And (6), of course is rejected in D.

However, this does not do justice to the example, for there is certainly a sense in which we may conclude from the facts of the case that Arthur is obligated to bandage the man whom Benjamin robs. It seems to me that this sense is exactly this: If Arthus is obligated to bandage Robert given that Robert is the man whom Benjamin robs, then Arthus is obligated to bandage the man whom Benjamin robs given that Robert is the man that Benjamin robs. This inference is sanctioned by the principle,

(7)   if $\vdash A \ \& B. \supset C$ then $\vdash O(A/B) \supset O(A \ \& \ C/B)$

which is correct, and accepted in our calculus (by AC 4 and RC 2). Leaving off the condition at any point destroys the validity of the inference, however; Arthur may, for example, have a primary obligation to help

Robert, but not have that obligation given that helping Robert will advance the cause of the Antichrist; and he may have an obligation to help a man who was robbed given that someone was robbed, but not *simpliciter*. Thus this example shows very clearly not the incorrectness of RC 2, but the inadequacy of the means of expression in *D*.

V. A SEMANTIC ACCOUNT OF CONDITIONAL OBLIGATION

In Section II, I already sketched an interpretation of conditional obligations. There I made the interpretation rather general by not assuming that among the values assigned to the physically accessible states there was a maximum. In formal semantics we prefer generality, of course, and I shall now further liberalize our notions by not assuming that each possible world is assigned one value, but rather that it is assigned a set of values. These values I will assume to be ordered linearly. The first generalization probably affects the logic (in that the insistence on a finite set of values would probably lose us compactness); I think that the second generalization does not. Finally I shall not assume that the ordering of the values and/or the assignment of values to worlds remains the same; in terms of Powers' thought-experiments, tomorrow the pay-off machine may have been reprogrammed. (If that is not allowed, we would need an extra axiom eg. the S_4 – like $O(B) \supset O(O(B))$.)

Thus we define a *C-model structure* (briefly, *C*-ms) as a quadruple $M = \langle K, V, R, f \rangle$ where

(1) K and V are non-empty sets

(2) R is a function with domain K and such that for each α in K, $R_\alpha = R(\alpha)$ is an asymmetric, transitive and connected relation on a non-empty subset of V, that is:
(a) If $R_\alpha(u, w,)$ then not $R_\alpha(w, u)$
(b) If $R_\alpha(u, w)$ and $R_\alpha(w, z)$ then $R_\alpha(u, z)$
(c) If $u \neq w$ then $R_\alpha(u, w)$ or $R_\alpha(w, u)$ for all u, w, z in the field of R_α (hereafter, V_α).

(3) f is a function with domain K such that for each α in K, $f_\alpha = f(\alpha)$ is a mapping of K into subsets of V_α, and such that $\bigcup \{f_\alpha(\beta) : \beta \in K\}$ is not empty.

We may read "$R_\alpha(v, u)$" as "v is greater than u for α" and "$f_\alpha(\beta)$" as "the set of values of β with respect to α."

These model structures can now be used to define truth conditions in a language with statements of conditional obligations. The language LC has as syntax

(1)   an infinite set of sentential parameters
(2)   the logical signs $\sim$, $\supset$, O, $/$, $)$, $($
(3)   a set of sentences defined by
        (a) sentential parameters are sentences
        (b) if A, B are sentences so are $\sim A$, $(A \supset B)$, $O(A/B)$.

Other connectives are defined in the usual way.

The semantics of LC is given by defining its *admissible valuations* to be exactly the mappings v_α such that α is a member of a set K and v is a valuation on a C-ms, $M = \langle K, V, R, f \rangle$, this latter notion being defined by:

(4)   A *valuation on* a C-ms $\langle K, V, R, f \rangle$ is a function v defined on K such that for each α in K, $v(\alpha) = v_\alpha$ satisfies:
        (a) v_α maps the sentences of LC into $\{T, F\}$
        (b) $v_\alpha(\sim A) = T$ iff $v_\alpha(A) = F$
        (c) $v_\alpha(A \supset B) = T$ iff $v_\alpha(A) = F$ or $v_\alpha(B) = T$
        (d) $v_\alpha(O(A/B)) = T$ iff there is an element β in K and an element w such that $v_\beta(A \& B) = T$, $w \in f_\alpha(\beta)$ and $R_\alpha(w, u)$ for every element u belonging to $f_\alpha(\gamma)$ for every γ such that $v_\gamma(\sim A \& B) = T$.

Since (d) is somewhat complex, we will rephrase it. Let $\beta R_\alpha \gamma$ mean that $f_\alpha(\beta)$ has a member u such that $R_\alpha(u, w)$ for each w in $f_\alpha(\gamma)$. Secondly, let $K_v(A) = \{\delta \in K : v_\delta(A) = T\}$. Thirdly, let us say that $K_v(A) R_\alpha K_v(B)$ exactly if $K_v(A)$ has a member β such that $\beta R_\alpha \gamma$ for each γ in $K_v(B)$. Omitting the subscript 'v' when the context prevents ambiguity, we can now reformulate (4) (d) as:

(4)(d') $v_\alpha(O(A/B) = T$ iff
            $K(A \& B) R_\alpha K(\sim A \& B)$.

It is helpful to note that if $A \Vdash B$ then $K_v(A) \subseteq K_v(B)$, and that $K_v(A \& B) = K_v(A) \cap K_v(B)$, $K_v(A \vee B) = K_v(A) \cup K_v(B)$, and $K_v(\sim A) = K - K_v(A)$.

THE LOGIC OF CONDITIONAL OBLIGATION 161

As corollaries to this we observe that if $\Vdash A$ then $K_v(A) = K$ and if $A \equiv B$ is a truth functional tautology then $K_v(A) = K_v(B)$.

Let us now examine the axioms and rules of system CD in the light of this semantic account. Two common expressions used in formal semantics are defined hereby:

> A is *valid* in LC ($\Vdash A$) exactly if every admissible valuation of LC satisfies A (i.e. assigns T to A). X *semantically entails* A in LC ($X \Vdash A$) exactly if every admissible valuation of LC which satisfies X (i.e. satisfies every member of set X) also satisfies A.

Are all theorems of CD valid (in LC)? That is, are all axioms valid and do all rules preserve this property?

That AC1 and RC1 are all right is clear. When a given valuation v_α is the only one under discussion, let us say that value *is in* A when it belongs to $\bigcup \{f_\alpha(\beta) : \beta \in K_v(A)\}$. For AC2, suppose there is a value in $A \,\&\, C$ that is higher any in $\sim A \,\&\, C$; then the converse cannot hold. Hence if $O(A/C)$ is true, $O(\sim A/C)$ is not. For AC3 we may disregard the conditionalization, since it is the same throughout (just assume that $K_v(C) = = K$). Suppose then there is a value w in $A \supset B$ higher than any in $A \,\&\, \sim B$, and a value u in A higher than any in $\sim A$. Now w must lie either in $\sim A$ or in B. If it lies in $\sim A$, then u is higher than w. So we have a value in $\sim A$ higher than any in $A \,\&\, \sim B$, but a value in A higher than any in $\sim A$ or in $A \,\&\, \sim B$. The latter value, w, must therefore lie in $A \,\&\, B$. If a value z lies in $\sim B$ it lies in $A \,\&\, \sim B$ or in $\sim A$. Hence w in $A \,\&\, B$, and hence in B, is higher than any value in $\sim B$. On the other hand, if w does not lie in $\sim A$, it lies in B, and hence in $A \,\&\, B$. If w is higher than or equal to u, then it is higher than any value in $A \,\&\, \sim B$ or in $\sim A$, and hence higher than any value in $\sim B$. If u is higher than w, it is higher than any value in $A \,\&\, \sim B$, and hence lies in $A \,\&\, B$, so there is a value in B higher than any in $A \,\&\, \sim B$ or in $\sim A$. As we see, all possibilities substantiate AC3.

For RC2, we note that if $\Vdash A \supset B$, then $K_v(A) \subseteq K_v(B)$, so $K_v(A \,\&\, C) \subseteq K_v(B \,\&\, C)$, while $K_v(\sim B \,\&\, C) \subseteq K_v(\sim A \,\&\, C)$. Thus a value in $A \,\&\, C$ higher than any in $\sim A \,\&\, C$ will at once be a value in $B \,\&\, C$ higher than any in $\sim B \,\&\, C$. For RC3 we invoke the special condition on f that $\bigcup \{f_\alpha(\beta) : \beta \in K\}$ is not empty; since this is exactly the set of values in A if $\Vdash A$, we clearly have $K_v(A \,\&\, A) \, R_\alpha K_v(\sim A \,\&\, A)$ in that case. RC4 is substantiated at once by the consideration that $K_v(C) = K_v(D)$ if $\Vdash C \equiv D$.

Of AC4–AC7 I will explicitly discuss only AC5. Recall that $B(A/B)$ was defined as $O(\sim B/A \vee B)$; is that a good definition? We want to have $v_\alpha(B(A/B)) = T$ exactly if $K_v(A) R_\alpha K_v(B)$. But the latter condition says that some value in A is higher than any in B; such a value *can only* lie in A & $\sim B$. Hence $K_v(A) R_\alpha K_v(B)$ exactly if $K_v(\sim B$ & $A) R_\alpha K_v(B)$. And that is the case exactly if $v_\alpha(O(\sim B/A \vee B)) = T$. Now we see therefore that AC5 is valid exactly if R_α is transitive; and it is.

VI. COMPLETENESS OF THE SYSTEM CD

The discussion of CD at the end of the preceding section showed, in effect, that CD is *sound* with respect to LC: if A can be deduced from premises X via system CD then $X \Vdash A$ in LC. Now I want to show that CD is (strongly) *complete* with respect to LC: if $X \Vdash A$ in LC, then A can be deduced from X via CD. As a preliminary, an indifference relation is defined, and a number of theorems proved concerning CD.

'$S(A/B)$' for '$\sim B(A/B)$ & $\sim B(B/A)$'
T5 $\vdash B(A/B) \supset \sim B(B/A)$

For $B(A/B) \equiv O(\sim B/A \vee B)$. Assuming both $B(A/B)$ and $B(B/A)$ we have $O(\sim B/A \vee B)$ & $O(\sim A/A \vee B)$, hence $O(\sim A$ & $\sim B/A \vee B)$, hence $O(\sim(A \vee B)/A \vee B)$ by T2. But $\sim O(\sim(A \vee B)/A \vee B)$ by T3.

T6 $\vdash B(A/B) \supset [B(B/C) \supset B(A/C)]$

This follows at once from AC5.

T7 $\vdash S(A/B) \supset [B(A/C) \supset B(B/C)]$
T8 $\vdash S(A/B) \supset [B(C/A) \supset B(C/B)]$

These follow from AC6 and AC7 respectively.

T9 $\vdash S(A/A)$

This follows from $\sim B(A/A) \equiv \sim O(\sim A/A)$, and T3.
The following two theorems are tautologies.

T10 $\vdash S(A/B) \supset S(B/A)$
T11 $\vdash S(A/B) \vee B(A/B) \vee B(B/A)$
T12 $\vdash S(A/B) \supset [S(B/C) \supset S(A/C)]$

For suppose $S(A/B)$ and either $B(A/C)$ or $B(C/A)$. Then either $B(B/C)$ or $B(C/B)$ follows by T7–8.

T13   If $\vdash A \supset B$ then $\vdash \sim B(A/B)$

For $B(A/B) \equiv O(\sim B/A \vee B) \equiv O(\sim B/B)$ if $A \vdash B$ (for then $\vdash B \equiv A \vee B$; RC4). But $\vdash \sim O(\sim B/B)$ by T3.
As corollary we have:

T14   If $\vdash A \equiv B$ then $\vdash S(A/B)$
T15 $\vdash B(B/\sim(A \supset A)) \vee S(B/\sim(A \supset A))$

Immediate from T13 and T11.

T16 $\vdash O(B/B) \equiv B(B/\sim(A \supset A))$

$B(B/\sim(A \supset A)) \equiv B(B/B \,\&\, \sim B) \equiv O(B \supset B/B)$. But $O(B/B) \supset O(B \supset B/B)$ by RC2.
Conversely, $O(B \supset B/B)$ implies $O(B/B)$ by AC4.

T17 $\vdash O(A/B) \supset O(A \,\&\, B/A \,\&\, B)$

Suppose $\sim O(A \,\&\, B/A \,\&\, B)$. Then by T15 and T16 we have $S(A \,\&\, B/\sim \sim(A \supset A))$. Now $S(A \,\&\, B/\sim(A \supset A))$ implies $\sim B(A \,\&\, B/\sim A \,\&\, B)$ by T15 and T7. But $O(A/B) \equiv B(A \,\&\, B/\sim A \,\&\, B)$ by T4.

T18 $\vdash \sim B(D/D \,\&\, C) \vee \sim B(D/D \,\&\, \sim C)$

Suppose $B(D/D \,\&\, C)$ & $B(D/D \,\&\, \sim C)$. Then $O(\sim(D \,\&\, C)/D)$ & $O(\sim(D \,\&\, C)/D)$. By T2 we arrive at $O(\sim D/D)$ which is impossible by T3.

T19 $\vdash B(A \,\&\, B/\sim(A \supset A)) \,\&\, S(\sim A \,\&\, B/\sim(A \supset A)). \supset$
         $\supset O(A/B)$

Assume the antecedent. By T8, $B(A \,\&\, B/\sim A \,\&\, B)$. Hence $O(A/B)$.

We can now turn to the completeness proof proper. Every set of sentences that is consistent with respect to CD can be extended into a maximal set, by Tukey's lemma and the fact that the deducibility relation is finitary. Hence it suffices to show that all maximal consistent sets can be satisfied. We shall make up a single C-ms to show this: $M = \langle \sum, V, R, f \rangle$ where $\sum$ is the family of maximal consistent sets, $V = \bigcup \{V_\alpha : \alpha \in \sum\}$, and V_α, R_α, and f are as we shall now define them.

Let α be in $\sum$. For any sentence A, we define $[A]_\alpha = \{B : S(A/B) \in \alpha\}$, and define $V_\alpha = \{[A]_\alpha : A$ is a sentence of LC$\}$. In addition, we define $R_\alpha = \{\langle[A]_\alpha, [B]_\alpha\rangle : B(A/B) \in \alpha\}$, and for β in $\sum$, define $f_\alpha(\beta) = \{[A]_\alpha : A \in \beta$ and for all B in β, $\sim B(B/A) \in \alpha\}$. We shall omit the subscripted Greek letters sometimes when confusion is prevented by context.

With respect to these definitions, we note first that the relation $S(A/B) \in \alpha$ between A and B is an equivalence relation (T9, 10, 12) and that if $S(A/B) \in \alpha$, then if $B(A/C) \in \alpha$, so is $B(B/C)$, and if $B(C/A) \in \alpha$, so is $B(C/B)$, by T7–8. Hence we deduce readily that R_α is transitive (T6), asymmetric (T5), and connected (T11), and has $[\sim(A \supset A)]_\alpha$ as lowest element (T15). By a 'lowest' element of a collection, we mean one which does not bear R_α to any (other) member of that collection. In fact, we may call $[\sim(A \supset A)]_\alpha$ the lowest element. For we have either $B(B/\sim \sim A \supset A))$ or $S(B/\sim(A \supset A))$: in the former case $[B]$ is higher, and in the latter case $[B] = [\sim(A \supset A)]$.

The relation R_α is therefore as the definition of C-ms requires. The lemma which follows and which will also be of use later on, shows that the requirement that $\bigcup \{f_\alpha(\beta) : \beta \in K\}$ be non-empty is also fulfilled, since $O(B \supset B/B \supset B) \in \alpha$ by RC3.

For a given sentence B, we are going to define a set $T_\alpha(B)$, which will be a maximal consistent set under suitable conditions. We define $T_\alpha(B)$ to be the deductive closure of the set $\{D_1, D_2, ...,\}$ where $C_1, C_2, ...$ are exactly the sentences of LC and

$D_1 = B$
$D_{i+1} = D_i \& C_i^*$ where C_i^* is C_i if
$\sim B(D_i/D_i \& C_i) \in \alpha$, and C_i^* is $\sim C_i$ otherwise.

Clearly, $T_\alpha(B)$ is maximal if consistent.

Lemma If $O(B/B) \in \alpha$, then $T_\alpha(B) \in \sum$
and $f_\alpha(T_\alpha(B)) = \{[B]_\alpha\}$

Proof. First, if $O(B/B) \in \alpha$, then $B(B/\sim(A \supset A)) \in \alpha$, by T16. Secondly, either $\sim B(D_i/D_i \& C_i)$ or $\sim B(D_i/D_i \& \sim C_i)$ by T18; hence $\sim B(D_i/D_{i+1})$. We can now conclude that for $J = 1, 2, 3, ..., [D_i]R_\alpha[\sim(A \supset A)]$, hence $T_\alpha(B)$ is consistent (see T14). We want to show that for all C in $T_\alpha(B)$, $\sim B(C/B) \in \alpha$. Well, if C is in $T_\alpha(B)$, then it logically implied by D_i for some index i. But then $\sim B(C/D) \in \alpha$ by T13, and since in addition $\sim B(D_i/B) \in \alpha$, because $\vdash D_i \supset B$ (see Theorem T13); we conclude

$\sim B(C/B) \in \alpha$ by AC6. So $[B]_\alpha$ is in $f_\alpha(T_\alpha(B))$; in addition, if $[C]_\alpha$ also belongs, we must have $\sim B(B/C)$ and $\sim B(C/B)$, hence $[B]_\alpha = [C]_\alpha$. This ends the proof.

Theorem There is a valuation v on $M = \langle \sum, V, R, f \rangle$ such that v_α satisfies α, for each α in $\sum$.

Proof. We define v by $v_\alpha(A) = T$ if $A \in \alpha$, and $v_\alpha(A) = F$ otherwise. That v satisfies clauses (a)–(c) in the definition of a valuation on a C-ms is obvious. To show that clause (d) is satisfied, we consider two cases.

Case 1. $O(A/B)$ is in α. Then $O((A \& B)/A \& B)) \in \alpha$ by T17. Hence $f_\alpha(T_\alpha(A \& B)) = \{[A \& B]\}$. Now suppose that $\sim A \& B \in \beta$, and $f_\alpha(\beta)$ contains $[C]$. Then $[C]$ can be no higher than $[\sim A \& B]$. But since $O(A/B) \in \alpha$, so is $B(A \& B/\sim A \& B)$, by T4. Thus $[A \& B]$ is higher than $[\sim A \& B]$ and hence higher than $[C]$. So there is an element in $A \& B$ which is higher than any in $\sim A \& B$, in our earlier terminology.

Case 2. $O(A/B) \in \alpha$. Consider $(\sim A \& B)$: either $O(\sim A \& B/\sim A \& B)$ is in α or it is not. If it is in α, then $f_\alpha(T_\alpha(\sim A \& B)) = \{[\sim A \& B]\}$, which contains a value no lower than $[A \& B]$, since $\sim O(A/B) \in \alpha$, by T4, and hence no lower than any value to be found in $f_\alpha(\beta)$ when β contains $A \& B$. Suppose that $O(\sim A \& B/\sim A \& B)$ is not in α. In that case $S(\sim A \& B/\sim(A \supset A)) \in \alpha$ by T15–16; we claim that similarly $S(A \& B/\sim \sim(A \supset A)) \in \alpha$, so that if β in $\sum$ contains $A \& B$ or $\sim A \& B$, then $f_\alpha(\beta) = \{[\sim(A \supset A)]_\gamma\}$. But our claim follows from T15 and T19, given that $O(A/B)$ is not in α. This ends the proof.

VII. THE ANDERSON MODIFICATION MODIFIED

A. R. Anderson introduced a device, since known as the *Anderson modification*, which was designed to reduce deontic logic to alethic modal logic.[14] We choose a normal system of modal logic, add a constant S (generally read as "All hell breaks loose."), the axiom $\vdash \Diamond \sim S$, and define the monadic operator O by $O(A) \equiv \Box(\sim A \supset S)$. Then all the laws of D are provable. (In addition, the semantics of modal and deontic logics developed since then shows that no non-theorems of D will be provable in M: we take S to be false exactly in the ideal possible worlds, true in the non-ideal ones.)

As a reduction in the technical sense, in which a system is classified as alethic or deontic by the syntactic form of its theorems, the Anderson modification is a successful and highly useful device. Anderson also introduced the thesis that this is a key to the correct interpretation of deontic concepts: there is something (which is bad) and which happens exactly if any obligation is violated. We do not have to look far to find what that something is, of course: what happens in all and only those cases in which an obligation is violated is that an obligation is violated. Critics have urged that this clearly does not explicate deontic concepts in non-deontic terms.[15] If we were to find, say, that physical laws are such that in any physically possible world, some obligation is violated if and only if someone proves that deontic logic is reducible to alethic logic, we could take S to state *that* : But it would still have to be admitted that the *meaning* of deontic terms cannot be given in terms of physical necessity and logical activity.

Be that as it may, Anderson's translation into modal logic shows very clearly the shallow character of deontic distinctions that can be expressed in D. Some moral violations lead to the Deluge, some to Hell, some to Purgatory, some to prison, some to gout, and some to gubernatorial disapproval; but in D they are classified alike as leading to something bad. In addition, one and the same course of action may earn time in prison and merit in heaven, and a choice may be between the devil and the deep blue sea; no such distinctions are possible in D.

For CD, the Anderson modification is not possible. For example, we cannot find one sentence true in every member of $K_v(A \& B)$ and false in every member of $K_v(\sim A \& B)$ whenever $O(A/B)$ is true; this would make the conjunction of $O(A/B)$ and $O(A \& C/B)$ impossible when $K_v(A)$ is not the same as $K_v(C)$. However, we can introduce a device similar to Anderson's. Let us suppose that when a world (situation, state of affairs) has some value, this is it because in it there exists something that has that value (with or without further qualifications, such as that there be nothing in that same world with lesser value). Then our interpretation, when used to answer someone who asks "What ought I to do?", is: if K is the set of possible outcomes you can achieve, then $O(A/B)$ exactly if there is something X such that $\Diamond (A \& B \& x$ exist) is true, and for all y such that $\Diamond (\sim A \& B \& y$ exists) is true, x has greater value than y (with or without qualifications, such as that y have some value).

The supposition made in the preceding paragraph is philosophically either dubious or unenlightening, while technically beyond reproach. For example, if the supposition is not true under any non-trivial interpretation, we can simply agree to count the value(s) of a world among its inhabitants, using perhaps a special predicate to express the distinction between them and normal residents (animate, inanimate, or abstract). Suitable reinterpretation (and perhaps relativization of the quantifiers) would help to keep the language's resources as great as before. The only problem I can see is that I am making it impossible for a valuable world to be empty. I am powerless to deal with that case unless an empty world is devoid of moral value. Barring recent progress in moral theory as yet unknown to me, I cannot find that admission very damaging.

As a convenient system of quantificational modal logic I shall choose Q_1M, a system devised by Thomason.[17] The adjustments I make are to ignore names and definite descriptions, and to stipulate that there be at least one monadic and one dyadic predicate. To be precise, the new syntax will contain all the sentence parameters of LC, and the logical signs $\Box$, $\sim$, $\supset$,), (, =, for each integer $n > 1$ a set of predicates (of degree n), these sets being non-empty for $n=1$ and $n=2$, an infinite set of variables, and the set of sentences is defined by induction as usual. We assume a well-ordering of the expressions, and designate the first monadic predicate as $E!$, the first dyadic predicate as R.

The axioms and rules for Q_1M are those for quantificational logic with identity, and in addition

AM 1 $\vdash \Box A \supset A$
AM 2 $\vdash \Box (A \supset B) \supset . \Box A \supset \Box B$
AM 3 $\vdash (x) \Box A \equiv \Box (x) A$
AM 4 $\vdash (x)(y)(x = y \supset \Box x = y)$
RM 1   If $\vdash A$ then $\vdash \Box A$

To form the extended system Q_1M^+ we add

AM 5 $\vdash (x)(y)(Rxy \supset \sim Ryx)$
AM 6 $\vdash (x)(y)(z)(Rxy \supset : Ryz \supset Rxz)$
AM 7 $\vdash (x)(y)(x \neq y \supset Rxy \vee Ryx)$
AM 8 $\vdash (Ex)(\Diamond E!x)$
Def.   '$O(A/B)$' for '$(Ex)(\Diamond (A \& B \& E!x) \& (y)(\Diamond (\sim A \& B \& Ey!) \supset Rxy))$'
       where x and y are distinct variables.

We shall read "Rxy" as "x is higher (in value) than y", and "$E!x$" as "x exists"; clearly we should read "$(x)A$" as "for all possibles x, A", since $(x)(E!x)$ cannot be deduced. A sentence is a theorem in $Q_1 M^+$ exactly if it can be derived in $Q_1 M$ from premises of the form given in AM 5-8, hence a reduction of CD to $Q_1 M^+$ is also a reduction to $Q_1 M$. We prove the reduction semantically, by adjusting the models of $Q_1 M$ so that they satisfy $Q_1 M^+$.

A *Q-ms* is a triple $<K, \Pi, D>$ where K and D are non-empty sets and Π a reflexive dyadic relation on K. A *valuation* v on $<K, \Pi, D>$ is a function which maps the variables into D and assigns to each member α of K a mapping v_α fulfilling the conditions

(a) $v_\alpha(x) = v(x)$ for any variable x
(b) $v_\alpha(A) \in \{T, F\}$ for any sentence parameter A
(c) $v_\alpha(P) \subseteq D^n$ for any n-ary predicate parameter P
(d) $v_\alpha(R)$ is asymmetric, transitive, and connected in D
(e) $\bigcup \{v_\beta(E!) : \alpha \Pi \beta\}$ is not empty

The truth-values of the sentences that are not sentence parameters are then given inductively:

(1) $v_\alpha(x=y) = T$ iff $v_\alpha(x) = v_\alpha(y)$
(2) $v_\alpha(Px_1 \ldots x_n) = T$ iff $<v_\alpha(x_1), \ldots, v_\alpha(x_n)> \in v_\alpha(P)$
(3) $v_\alpha(\sim A) = T$ iff $v_\alpha(A) = F$
(4) $v_\alpha(A \supset B) = T$ iff $v_\alpha(A) = F$ or $v_\alpha(B) = T$
(5) $v_\alpha((x)A) = T$ iff $v^d/x_\alpha(A) = T$ for each d in D (where v^d/x is exactly like v except for assigning d to x)
(6) $v_\alpha(\Box A) = T$ if $v_\beta(A) = T$ for each β in such that $\alpha \Pi \beta$.
(7) $v_\alpha(A) = F$ if $v_\alpha(A) \neq T$.

An admissible valuation of this syntax is a mapping v_α ahere α is a member of K and v a valuation on $<K, \Pi, D>$ for some *Q-ms* $<K, \Pi, D>$. The completeness theorem for $Q_1 M^+$ with respect to LC^+ so formed is a corollary to Thomason's completeness proof for $Q_1 M$.

Suppose now that $M = <K, \Pi, D>$ is a *Q-ms* and v a valuation on M. We define the structure $M^1 = <K, V, R, f>$ as follows: $V = D$, $R_\alpha = v_\alpha(R)$, $f_\alpha(\beta) = \Lambda$ if not $\alpha \Pi \beta$, otherwise $f_\alpha(\beta) = \{d \in D : d \in v_\beta(E!)\}$. We note that $V_\alpha = V$ for all α, R_α is antisymmetric, transitive, and connected in V_α, and $\bigcup \{f_\alpha(\beta) : \beta \in K\}$ is not empty. Hence M^1 is a *C-ms*. If we look only at sentences formed from sentence parameters and $\sim$, &, O, then v obviously fulfills conditions (a)–(c) in the definition of a valuation on a C-ms. To show that it also satisfies (d), we need only show that $\bigcup \{f_\alpha(\beta): v_\beta(A)=T\}$, i.e. the set of values in members of $K_v(A)$, is just $\{d \in D: v^d/x_\alpha (\Diamond A \ \& \ E!x))\}$. Well d is in the first set exactly if $v^d/x_\beta(E!x) = T$ for some β such that $\alpha \Pi \beta$ and $v_\beta(A) = T$. But that is the case exactly if $v^d/x_\alpha(\Diamond (A \ \& \ E!x)) = T$.

Thus admissible valuations of LC^+ are also admissible valuations of LC. To prove the converse, suppose that $M^1 = <K, V, R, f>$ is a C-ms, and α a specific member of K. We define the structure $M = <K, \Pi, D>$ as follows: $\Pi = K^2$, $D = V_\alpha$. Clearly M is a Q-ms. Now let v^1 be a valuation of M^1; we define a valuation v on M, with the hope of showing that v_α^1 and v_α are the same as far as the sentences of LC are concerned.

For any β, and any sentence parameter A, let $v_\beta(A) = v_\beta^1(A)$. Let $v_\beta(E!) = f_\alpha(\beta)$, and let $v_\beta(R) = R_\alpha$. What v does for variables and predicates other than $E!$ and R is immaterial: we assume a suitable choice is made. Then v is a valuation on M, since R_α has the requisite properties and $\bigcup \{f_\alpha(\beta) : \beta \in K\} = \bigcup \{v_\beta(E!) : \alpha \Pi \beta\}$ is not empty. We must show that for the LC sentences, $v_\alpha^1(O(A/B)) = v_\alpha(O(A/B))$. For this we note first that $\bigcup \{f_\alpha(\beta) : v_\beta(A \ \& \ B) = T\} = \{d \in D : d \in v_\beta(E!x \ \& \ A \ \& \ B)$ for some β such that $\alpha \Pi \beta\} = = \{d \in D : v^d/x_\alpha(\Diamond(A \ \& \ B \ \& \ E!x)) = T\}$; $\bigcup\{f_\alpha(\gamma) : v_\gamma(\sim A \ \& \ B) = T\} = \{e \in D : v^e/y_\alpha(\Diamond(\sim A \ \& \ B \ \& \ E!y)) = T\}$. Thus $v_\alpha((E!x)(\Diamond(A \ \& \ B \ \& \ E!x)\&(y)(\Diamond(\sim A \ \& \ B \ \& \ E!y) \supset Rxy)) = T$ exactly if there is a member d of the first set that bears R_α to each member of the second set. So every admissible valuation of LC can be extended into an admissible valuation of LC^+.

VIII. SCEPTICAL POSTSCRIPT

At several points in the preceding sections I argued that there are moral distinctions that simply cannot be expressed adequately in the language of absolute obligations. I hold exactly the same view concerning the language of conditional obligations just constructed, even though I think it is an improvement with respect to the problems considered so far.

First, in constructing CD I decided to accept von Wright's criterion, that $O(A/B \supset B)$ should follow that logic of absolute obligation which is now standard in deontic logic. This made it necessary to assume that values are ordered linearly. This means in turn that $O(A/C)$ and $O(\sim A/C)$ can not both be true. However, we often find ourselves in a situation in which we have at least *prima facie* conflicting obligations. Hence it would be more apt to say that we have here a logic of obligations that remain after obligational conflicts are resolved.

If we hold that 'ought implies can', at least in the logical sense of 'can' and theorem T2 holds, then it follows that there are no moral conflicts incapable of resolution, that is, no possible situation in which we really (and not just *prima facie*) have obligations that cannot possible all be fulfilled. This is the point of view of, for example, Castañeda: "it is the function of the ethical 'ought' and the *ethos* it governs to solve the conflicts of duties... "[18]. I would not deny that *ideally* this is the case, but I do not believe that it is true of actual morality. If a legal system leaves problems unsolved, or has laws that conflict in a given unforeseen situation, the judicial system amends the law by precedent and reinterpretation. But there is no judical system for morality, and new moral rules do not come into existence by fiat or plebiscite.

A second point of criticism concern the formula $O(B/B)$.[19] This is almost always true; it is true if some value attaches to some possible world (attainable outcome) in which B is true. That means then that the violation of a (primary) obligation is a (secondary) obligation relative to the assumption that the obligation is in fact violated. 'Rightly understood' of course, it is true; if we have put ourselves in a situation in which a certain ideal can no longer be attained, then doing the best one can will involve not attaining that ideal. No use crying over spilt milk. But clearly there are also many moral evaluations and value judgments concerning such a situation which our schema leaves out of account altogether. The moral questions that can be asked go far beyond the simple "What ought to be done now?" Perhaps the addition of tense-logical machinery will alleviate some of the shortcomings, by allowing answers also to "What ought to have been done?"[20]

Finally, in moral discourse obligation is qualified and relativized in many ways. In the article cited above, Castañeda argued that "ought" should be subscripted, to allow the expression of obligations due to etiquette, laws, professional standards, and so on. These are non-ethical obligations. But ethical obligations might be divided into sub-categories too. There are duties to one-self, and duties to others; obligations incurred by promises and obligations incurred by acquiescence; obligations devolving upon one due to one's rank or due to one's happening to be in a certain place, and so on. Each of these can certainly be subdivided into sub-sub-categories: duties to others may comprise conflicting duties toward Peter and toward Paul; conflicting obligations might be incurred, perhaps quite unintentionally, by promises to Mary and to Martha, and so on.

This seems to me to be a case of 'variable polyadicity', and the use of subscripts can hardly be adequate to handle it, though it provides a first approximation. One avenue to approach might be to introduce prepositions and other adverbial devices into LC^+, and to consider the effects of adverbial modifications of the predicate R there. But the success of any such attempt would require the previous success of a systematic analysis of rights, duties, values, and obligations.

University of Toronto, Toronto

NOTES

* The research for this paper was supported by Canada Council grant 69-0650. The author wishes to acknowledge gratefully his debt to A. al-Hibri, 'A Critical Survey in Deontic Logic'. In forthcoming work on counterfactuals, David Lewis discusses the relation of the system CD and language LC here constructed to this minimal logic of counterfactuals CO and to the work of Bengt Hanson, 'An Analysis of Some Deontic Logics'.

[1] L. Åqvist, 'Good Samaritans, Contrary-to-Duty Imperatives, and Epistemic Obligations'.
[2] F. Fitch, 'Natural Deduction Rules for Obligation'.
[3] While the paradox was already presented by Bradley, it is related to deontic logic by R. M. Chisholm, in 'Contrary-to-Duty Imperatives and Deontic Logic'.
[4] G. H. von Wright, 'A Note on Deontic Logic and Derived Obligation'.
[5] L. Powers, 'Some Deontic Logicians'.
[6] al-Hibri, *op. cit.*, p. 32.
[7] G. H. von Wright, 'An Essay in Deontic Logic and the General Theory of Action', especially pp. 25 and 35.
[8] In Section IV, I shall discuss an argument by Castañeda against a similar principle.
[9] L. Åqvist, 'Deontic Logic Based on a Logic of 'Better''.
[10] The concept of conditional permission called 'natural' by von Wright (*op. cit.*, p. 35) which apparently fits N. Rescher, 'An Axiom System for Deontic Logic' has as theorem $O(A/B \lor C) \equiv O(A/B) \lor O(A/C)$, if "$O($" is defined as "$\sim P) \sim$" as usual. But on our interpretation even the weaker $O(A/B) \supset O(A/B \lor C)$ does not hold. For to steal and make restitution is better than to steal and not make restitution, but to have a cup of tea instead is better yet.
[11] von Wright, *op. cit.* p. 30.
[12] *op. cit.*, pp. 371-3.
[13] H.-N. Castañeda, 'Acts, the Logic of Obligation, and Deontic Calculi', especially pp. 13-4.

[14] A. R. Anderson, 'A Reduction of Deontic Logic to Alethic Modal Logic'.
[15] See for example J. Berg, 'A Note on Deontic Logic' and P. H. Nowell-Smith and E. J. Lemmon, 'Escapism: the Logical Basis of Ethics'.
[16] This makes the point (cf. Lemmon and Nowell-Smith, *op. cit.*, p. 291) that violations are not always followed by appropriate sanctions simply irrelevant.
[17] See R. H. Thomason, 'Some Completeness Results for Modal Predicate Calculi', and 'Modal Logic and Metaphysics'.
[18] H. N. Castañeda, 'A Theory of Morality', p. 345.
[19] For this paragraph I am indebted to the critical discussion of Åqvist's system DL_w in al-Hibri, *op. cit.*, pp. 36–7.
[20] Cf. R. H. Thomason, 'Deontic Logic as Founded on Tense Logic'.

BIBLIOGRAPHY

al-Hibri, A., 1968, 'A Critical Survey in Deontic Logic', mimeographed, Wayne State University.
Anderson, A. R., 1958, 'A Reduction of Deontic Logic to Alethic Modal Logic', *Mind* **67**, 100–3.
Åqvist, L., 1963, 'Deontic Logic Based on a Logic of 'Better'', *Acta Philosophica Fennica* **16**, 285–90.
Åqvist, L., 1967, 'Good Samaritans, Contrary-to-Duty Imperatives, and Epistemic Obligations', *Nous* **1**, 361–79.
Berg, J., 1960, 'A Note on Deontic Logic', *Mind* **69**, 566–7.
Castañeda, H.-N., 1957, 'A Theory of Morality', *Philosophy and Phenomenological Research* **17**, 345.
Castañeda, H.-N., 1968, 'Acts, the Logic of Obligation, and Deontic Calculi', *Philosophical Studies* **19**, 13–26.
Chisholm, R., 1963, 'Contrary-to-Duty Imperatives and Deontic Logic', *Analysis* **24**, 33–6.
Fitch, F., 1966, 'Natural Deduction Rules for Obligation', *American Philosophical Quarterly* **3**, 27–38.
Hanson, B., 1969, 'An Analysis of Some Deontic Logics', *Nous* **3**, 373–98.
Lambert, K. (ed.), 1969, *The Logical Way of Doing Things*, Yale University Press, New Haven.
Lambert, K. (ed.), 1970, *Philosophical Problems in Logic*, Reidel Publishing Company, Dordrecht.
Nowell-Smith, P. H. and Lemmon, E. J., 1960, 'Escapism: the Logical Basis of Ethics', *Mind* **69**, 289–300.
Powers, L., 1967, 'Some Deontic Logicians', *Nous* **1**, 381–400.
Rescher, N., 1958, 'An Axiom System for Deontic Logic', *Philosophical Studies* **9**, 24–30.
Thomason, R. H., 1969, in K. Lambert (ed.), *The Logical Way of Doing Things*, pp. 119–46.
Thomason, R. H., 1970a, in K. Lambert (ed.), *Philosophical Problems in Logic*, pp. 56–76.
Thomason, R. H., 1970b, 'Deontic Logic as Founded on Tense Logic', presented at Temple University Conference on Deviant Semantics, Dec. 1970.
von Wright, H., 1956, 'A Note on Deontic Logic and Derived Obligation', *Mind* **65**, 507–9.
von Wright, H., 1968, 'An Essay in Deontic Logic and the General Theory of Action', *Acta Philosophica Fennica* **21**.

HARRY BEATTY

ON EVALUATING DEONTIC LOGICS
Comments on van Fraassen's Paper

Professor van Fraassen, in his paper 'The Logic of Conditional Obligation', argues that certain problems, raised for the minimal deontic logic D by contrary-to-duty imperatives, by the Good Samaritan paradox, and by Powers' John and Suzy paradox, can be handled by his conditional logic of obligation CD. In my comments I shall for the most part neither attack nor support this claim: rather, I shall point out some reasons why I find it difficult to evaluate.

I shall begin with some general reasons why I have problems evaluating arguments for and against deontic systems. Typically such arguments turn on appeals to features of natural language captured or not captured by the deontic system in question. Take for example the specific argument constructed by Åqvist against D based on contrary-to-duty imperatives. He argues (Åqvist, 1967, p. 364) that the following set of sentences is intuitively consistent and independent (i.e. none of the four can be inferred from the other three), but that the sentences cannot be formalized in D so as to be consistent and independent.

(I)   It ought to be that Smith refrains from robbing Jones.
(II) Smith robs Jones.
(III) If Smith robs Jones, he ought to be punished for robbery.
(IV) It ought to be that if Smith refrains from robbing Jones he is not punished for robbery.

For example, if we paraphrase (I)–(IV) into D as

(1) $O(\sim A)$
(2) A
(3) $A \supset O(B)$
(4) $O(\sim A \supset \sim B)$

respectively, we can derive a contradiction. For $O(B)$ follows from (2) and (3), while $O(\sim B)$ follows from (1) and (4), since D allows distribution of 'O' across '$\supset$'. Similar problems arise for other likely paraphrases.

M. Bunge (ed.), Exact Philosophy, 173–178. All Rights Reserved
Copyright © 1973 by D. Reidel Publishing Company, Dordrecht-Holland

One obvious difficulty with this example is that, read literally, (IV) expresses the strange moral belief that Smith should be punished for robbery only if he robs *Jones*, not if he robs someone else. But perhaps the force of the example could be retained if (IV) were replaced with a sentence expressing a more plausible moral belief. In any case, Åqvist's example is difficult to evaluate for other reasons, which I can only indicate briefly here.

To begin with, both (III) and (IV) contain the troublesome little word 'if'. The example assumes that both of these sentences can be translated adequately using material implication. But this assumption can be questioned easily enough.

Secondly, if we accept the dictum that 'ought' implies 'can', it seems that a sentence like (I) should be paraphrased so as to imply a paraphrase of "Smith can refrain from robbing Jones". A similar remark applies to (III) and (IV). Perhaps the example points in some way to a failure to integrate deontic logic with some kind of alethic logic, rather than to any problem with the particular deontic system D.

Finally, this example may point to the need to incorporate some kind of tense logic into any system of deontic logic. (III) means that Smith ought to be punished *after* he robs Jones, but this is not indicated in any way in the translation into D. Further, it may be argued that (I) and (II) cannot be true at the same time. (I), it seems, can be true only *before* (II) becomes true. Again some incorporation of tense logic seems to be indicated.

No doubt other difficulties of this kind could be added to the list. But these difficulties raised by Åqvist's example will suffice to illustrate my general position on the evaluation of deontic systems, that evaluation of such system by appeal to ordinary linguistic intuition is at present next to impossible. Take any simple set of sentences which represents some moral beliefs and relevant facts. Adequate formalization of such a set of sentences typically requires, not only formal devices to deal with expressions such as 'ought' and 'permitted', but also formal devices to deal with: senses of 'If ... then' other than material implication; temporal concepts; the alethic modalities; quantification into different kinds of contexts; comparatives; adverbs; and so on. Further, it requires that these devices be combined in a single system. Until many prior problems of formalization of natural language have been adequately solved, evaluation of deontic systems by appeal to natural language will be a tenuous business at best.

Having outlined my general position on the evaluation of deontic systems, I shall now turn to the particular difficulties connected with evaluating the system *CD*. Some difficulties of this kind are indicated by Professor van Fraassen himself in his paper, particularly in the final section where he indicates certain problems left unsolved by *CD*. In part my remarks about *CD* will be complementary to his: at other places I shall take issue with his discussion. My comments will be divided into (i) remarks on the rule RC 2 of *CD* and (ii) remarks on the John and Suzy paradox.

(i) *Doubts about RC 2*. In his paper, Professor van Fraassen is concerned to defend the following primitive rule of inference of *CD*.

RC 2   if ⊢ $A \supset B$ then ⊢ $O(A/C) \supset O(B/C)$

The reason he feels it requires defense is that it is analogous to the derived rule of *D* which says that if ⊢ $A \supset B$, then ⊢ $O(A) \supset O(B)$, and this latter rule of *D* is often thought to be the source of difficulties connected with Good Samaritan paradox and with contrary-to-duty imperatives. It also seems from his discussion to be connected with the John and Suzy paradox.

Let us look more closely at what RC 2 involves. *Prima facie*, what it appears to say is that you are obligated to do (given some condition *C*) all the logical consequences of what you are obligated to do (given *C*). And put this way, it seems intuitively unobjectionable. A second look, however, may raise some doubts. Consider the following simple puzzle, known in the literature on deontic logic as Ross' paradox.

RC 2 clearly licenses the inference of $O(A \vee B/C)$ from $O(A/C)$. Now the following situation might be one in which a sentence of the form $O(A/C)$ is true while $O(A \vee B/C)$ is false. Tom has borrowed a rake from Albert, the time for which he borrowed it is up, and he ought to take it back. Thus the following sentence is to be accepted as true:

(5)   Tom ought to give the rake to Albert, given that Tom borrowed the rake form Albert.

On the other hand, there is an argument against accepting

(6)   Tom ought to give the rake to Albert or give the rake to Bill given that Tom borrowed the rake from Albert.

as true in the situation described. For (6) carries the suggestion that Tom can discharge his obligation by giving the rake to Bill, even if Bill is someone with no legitimate claim on the rake.

It is difficult to account in any precise way for the intuition that (6) is not true in the situation under consideration. I think, however, that the following can be accepted as the beginning of an explanation. In the first clause of (6), the words "Tom ought to" indicate that Tom has an obligation, while the words "give the rake to Albert or give the rake to Bill" give the content of the obligation by *describing the action* which Tom is said to be obligated to do. Only, in the situation under consideration, "give the rake to Albert or give the rake to Bill" *misdescribes* Tom's obligation, while "give the rake to Albert" describes it correctly. If we are asked what it is that Tom is obligated to do, "Give the rake to Albert or give the rake to Bill" is a wrong (or at best misleading) answer.

On this construal of obligation sentences, if (6) is rendered as $O(A \vee B/C)$, for example, the sentence letters 'A' and 'B' as well as the formula '$A \vee B$' should be regarded as *descriptions of actions*. Similarly for other sentence letters occurring within the scope of 'O' and before '/'. This new construal of sentence letters and formulas, however, places the rule RC 2 in a somewhat different perspective. Originally we viewed RC 2 as saying that you are obligated to do (given C) all the logical consequences of what you are obligated to do (given C). But now it can be viewed as saying that you can correctly be *described* as obligated to do (given C) all the logical consequences of what you can correctly be *described* as obligated to do (given C). Put this way, in terms of description, the rule seems much less plausible.

It should be clear from my preliminary remarks that I do not consider this a decisive objection to RC 2. I have pointed, I think, to a semantic feature of obligation sentences not dealt with by *CD*, and which furthermore *can't* be dealt with by *CD* because of the presence of RC 2. Whether theoretical considerations might *justify* overlooking this feature, what I have characterized as the descriptive aspect of obligation sentences, is of course another question. To my mind, this descriptive aspect is one of the most important and interesting semantic features of obligation language, and so I would want any theory of such language to account for it. But others, of course, will have different ideas as to what is important or interesting.

In any case, my discussion at least points to a general difficulty in testing deontic logics against natural language, even if in the long run this difficulty can be overcome. The difficulty is basically that, since obligation sentences in natural language contain descriptions of actions, sentence letters in deontic logic seem to have two roles to play – describers of actions and bearers of truth values. I, for one, have difficulty in keeping these two roles straight in considering arguments for and against deontic systems.

(ii) *John and Suzy*. Professor van Fraassen's discussion of Powers' John and Suzy paradox raises a number of puzzling questions. The main question raised is this: Does the paradox depend on any features of language peculiar to moral or normative discourse, or is it simply an example of a more general kind of difficulty that arises with respect to the analysis of natural language? I am inclined to take the latter view, that essentially the same problem arises in the analysis of non-normative language.

Consider the following pair of sentences.

(7)   If that match is struck, it will light.
(8)   If that match is struck and no oxygen is present, it will light.

It seems reasonable that, under appropriate circumstances, someone would want to hold that (7) is true but (8) is false. There is, however, no straightforward analysis in ordinary logic which allows for this possibility. For if we paraphrase (7) and (8) as implications, it is easy to see that the paraphrase of (7) will entail the paraphrase of (8).

Now I think it is clear that (7) and (8) present much the same difficulty as the John and Suzy paradox. A natural question to ask then is whether the semantics for LC could be adapted to deal with the non-normative case as well. This seems to me a reasonable line to investigate. After all, many problems in adequately representing ordinary language in logical systems arise from the divergence between conditionalization in ordinary language and material implication. But in *CD*, conditionalization is a logical primitive, distinct from material implication. Perhaps such a logical primitive is necessary if we are to better render natural language into a formal system.[1]

Of course, if we are to adapt the semantics of *CD* to the more general case, we must reinterpret the 'values'. A natural suggestion would be to treat them as probabilities, but no doubt further investigation would reveal other interpretations.

McGill University, Montreal

NOTE

[1] Professor van Fraassen, in the discussion following the reading of his paper at the Symposium in Exact Philosophy, supported and illustrated the view which I expressed in this paragraph.

BIBLIOGRAPHY

Åqvist, L., 1967, 'Good Samaritans, Contrary-to-Duty Imperatives, and Epistemic Obligations', *Nous* **1**, 361–79.

PART VIII

LEGAL PHILOSOPHY

CARLOS E. ALCHOURRÓN

THE INTUITIVE BACKGROUND OF NORMATIVE LEGAL DISCOURSE AND ITS FORMALIZATION

One of the motivations for the construction of a great many deontic logics has been the discovery of purely formal analogies with other modal logics. It is a well known fact that the first systems of deontic logic (von Wright, 1951a and 1951b) originated in the observation of strong analogies between the deontic operators P and O and the modal alethic operators of possibility and necessity (M and N), and the same is true for many later systems. But deontic logic may also be viewed as a rational reconstruction of the main features of norms and certain normative concepts, typical of the normative (legal and moral) discourse. On this view, a formal system must be supplemented by a sound interpretational analysis in order to make it clear what concepts are the explicanda of such a rational reconstruction. The clarification of the explicanda provides the adequacy criteria for the different systems of deontic logic. This does not mean that when confronted with two rival systems, in the sense that in one of them there is a principle (axiom, definition, rule of inference) which does not occur in the other, we must reject one of the systems. It may well be the case that, instead of being two different reconstructions of the same explicandum, they attempt to reconstruct different concepts hidden behind the ambiguity of the normative discourse. But in any case what is important is to delimit their areas of application and so to clarify as far as possible the corresponding explicanda.
In this paper I intend to do the following:

(1) I shall present several divergent systems of deontic logic well known in the literature, in order to analyze some differences that are motivated by tacit references to different explicanda.

(2) I shall try to show how the apparently opposed systems can be integrated when some very common usages of the normative concepts in the field of positive law and jurisprudence are taken to be the intuitive framework for the identification of the corresponding explicanda.

I

We define some deontic systems based on a denumerably infinite list of propositional variables, having $\supset$ (material implication), $\sim$ (negation), O (the O-deontic operator) and P (the P-deontic operator) as undefined connectives. The rules for well-formed formulas and the definitions of other propositional connectives are the usual ones. The axiom schemata and rules of inference are:

A0	$\vdash PA \equiv\, \sim O \sim A$	A0.1	$\vdash PA \supset\, \sim O \sim A$
A1	$\vdash (OA.OB) \supset O(A.B)$	A1.1	$\vdash P(A \vee B) \supset (PA \vee PB)$
A2	$\vdash OA \supset\, \sim O \sim A$	A2.1	$\vdash OA \supset PA$
A3	$\vdash O(OA \supset A)$		
A4	$\vdash O(A \supset OPA)$		
A5	$\vdash OA \supset OOA$		
A6	$\vdash PA \supset OPA$		

R1 If A is a tautology, then $\vdash A$
R2 If $\vdash A$ and $\vdash A \supset B$, then $\vdash B$
R3 If $\vdash A$, then $\vdash OA$
R4 If $\vdash A \supset B$, then $\vdash OA \supset OB$
R5 If $\vdash A \supset B$, then $\vdash PA \supset PB$

In terms of the axiom schemata and rules of inference we define the following systems:

$F = \{$A0–A1, R1–3$\} = [T(C)]$ $C2 = \{$A0–A1, R1, R2, R4$\}$
$D = \{$A0–A2, R1–3$\} = [T(D)]$ $D2 = \{$A0–A2, R1, R2, R4$\}$
$DM = \{$A0–A3, R1–3$\}$ $W = $ A01, A1, A1.1, R1, R2, R4, R5$\}$

$DB = \{$A0–A4, R1–3$\}$
$DS4 = \{$A0–A3, A5, R1–3$\}$
$DS5 = \{$A0–A3, A5, A6, R1–3$\}$

(The names between brackets are Lemmon's, 1966.)

The systems of the left column are substantially those contained in Hanson (1965), though I have slightly changed the presentation. A0 occurs usually as a definition and not as an axiom; Hanson and Lemmon use $O(A \supset B) \supset (OA \supset OB)$ instead of A1, but they are deductively equivalent in the presence of R4, which, of course, belongs to all systems; so I have chosen von Wright's style for the axioms of distributivity. The first two systems of the right column come from Lemmon (1966). (D2), with some restrictions in the formation rules, is von Wright's original system. W is the weak monadic P_1-O_1 calculus contained in von Wright (1967).

What I want to discuss in this paper is related to the domain of applicability of these systems, i.e. their intuitive background or interpretation. As will be obvious, when I refer in this paper to the interpretation of a system, I am not thinking of something in the style of Kripke's semantic models or McKinsey and Tarski's algebraic interpretations. I have chosen Hanson's and Lemmons' systems precisely because the problems of their semantic and algebraic interpretations are already solved.

The relations between the systems of deontic logic and the modal alethic systems of Lewis $S2$, $S4$, $S5$, M [T] of von Wright [Feys] and B (Brouwer) of Kripke are depicted in the following containment chart:

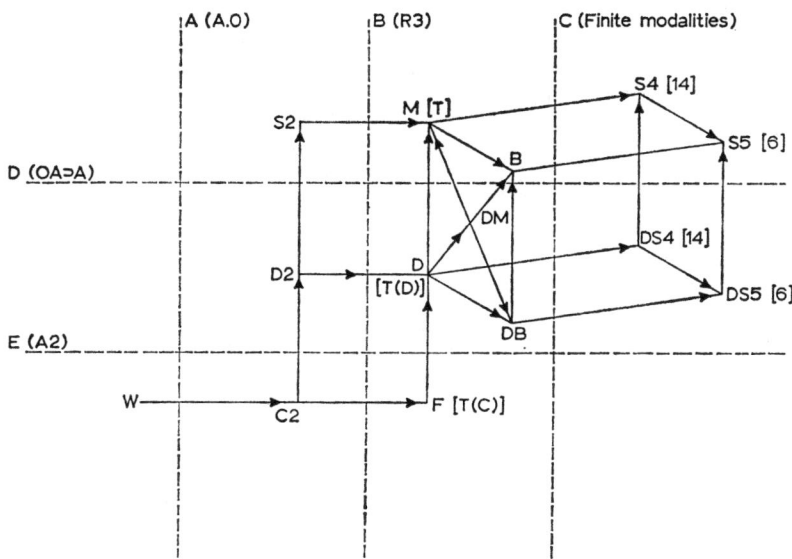

An arrow goes from a system to the one that properly contains it. Systems to the right of line A have the interdefinability axiom A0. Systems to the right of line B have the necessitation rule R3. Systems to the right of line C have a finite number of distinct modalities. The number in brackets is the corresponding number of modalities, as a consequence of the reduction axioms A5 and A6. Systems above line D have the alethic subordination axiom $OA \supset A$. Systems above Line E have the deontic subordination axiom A2.

I will examine some of the differentiating principles indicated by lines A, B, C, D and E of the preceding chart.

The deontic systems to the right of line C, i.e., $DS4$ and $DS5$, have been created following purely formal analogies. The problem was to obtain a deontic logic with the same 14 modalities of $S4$ and with the same 6 modalities of $S5$. The same is true of DB. In this sense the characteristic axioms of these systems – A4, A5 and A6 – require an intuitive interpretative foundation. I shall not discuss this question for the moment, but I want to point out that in some very important interpretations of the deontic operators the expressions that contain a deontic operator within the scope of another deontic operator are meaningless, so that it is convenient to consider them as ill-formed.

What has been regarded as the most characteristic feature of deontic logic is the acceptance of the deontic subordination axiom A2 ($OA \supset {\sim}O{\sim}A$) and the rejection of the alethic one ($OA \supset A$). That is why the systems between lines D and E have been called deontic in the literature.

Nevertheless, some authors have thought that this axiom does not express a necessary – a priori – conceptual truth, so that it should be eliminated from deontic systems. This is the opinion of Lemmon in his paper (1965), where he recommends the replacement of $D2$ by C. (Later on we shall see why he rejects also the systems at the right of line B.) Stenius, in his paper of 1963, favors the system $F\,[T(C)]$ over $D\,[T(D)]$. Von Wright also has expressed some doubts about the possibility of a purely conceptual justification of this principle.

I mention these different and sometimes vacillating opinions in order to stress the necessity of fixing clearly the intuitive framework that would help us to make a decision in these dubious cases.

II

Let us begin by considering a usual reading of the deontic operators. We shall read 'Op' as 'It is obligatory that p', or perhaps better as 'It is obligatory to see to it that p', '$O \sim p$' as 'It is prohibited (forbidden) that p' or 'It is obligatory that not p' and 'Pp' as 'It is permitted that p'. And let us direct our attention to a very common use of such phrases in lawyers' language, when they refer to enacted positive law. In order to simplify the matter we shall suppose – as does Hart (1961) in his book – that there is a population living in a territory in which an absolute monarch (Rex) reigns in such a way that he is unanimously recognized as the only source of law. Lawyers would say that it is obligatory that p when and only when Rex has commanded that p and they would also admit that Rex has commanded that p not only when he has directly referred to p in the dictum of his command, but also when the commanding of p may be inferred from his explicit orders. They would also say that it is forbidden that p when and only when Rex has commanded that not-p be the case. In this way they would say that it is obligatory and forbidden that p when Rex has directly or inderectly commanded that p and also that not-p. Of course in such a situation the subjects of the law would feel bewildered, because whatever they do in relation to p will be a transgression of some of Rex's orders. Naturally this is a regrettable situation but unfortunately not an impossible one, and lawyers know how frequently it occurs in daily life. In this sense, at least, it is perfectly clear that axiom A2 must be rejected as a law of the logic of obligation.

But when would lawyers say that it is permitted that p? Here things are not so clear, because 'permitted' has a fluctuating meaning. They would recognize that when Rex issues a law, the content of his dictum is not always a command; sometimes the law issued by Rex is an authorization to carry out some action. This kind of law does not give rise to an obligation nor to a prohibition but only to a permission, and in this sense they would say that it is permitted that p when and only when Rex has authorized p directly or indirectly. In this sense Rex's pronouncement authorizing p is necessary for the truth of 'It is permitted that p', so that the fact that Rex has not commanded that not p is not sufficient for the permissiveness of p. An argument of this kind is perhaps the one that lies behind von Wright's rejection of one half of Axiom 0, i.e., the implication $\sim O \sim A \supset PA$. But surely in this sense we must reject also A0.1, i.e., the other half of $A0$, because it may happen that Rex has issued a law authorizing p directly or indirectly and also a law commanding that not p; this, once again, would be a regrettable situation but not an impossible one.

In this sense I consider that A1.1 must also be rejected; the reasons for this will be given later.

If A0 and A2 are accepted in deontic logic, A2.1 is also a logical law, because it is an immediate consequence of both, but the elimination of A0 and A2 from deontic logic does not compel us to eliminate A2.1. I think that the latter is a law of deontic logic when the operators are interpreted in the indicated way because, when Rex has commanded that p, he has also indirectly authorized the doing of p, so when it is obligatory that p it is also permitted that p. Precisely because things are like that, one can find a better explanation of the bewilderment experienced when it is obligatory that p and not p for, when this is true, what happens is that something is permitted that is also prohibited, and, of course, this is perfectly possible as far as Rex may authorize p and may also command that not p.

But sometimes 'permitted' is not used by lawyers in the explained way. Many times they will identify p as permitted when and only when Rex has not commanded that not p, i.e., when it is not true that it is prohibited that p. I shall call *weak permission* this new sense of the *P*-operator and *strong permission* the preceding one. In the weak sense of P axiom A0 is, indeed, a necessary logical truth and it may be considered a definition. This is the only sense which those authors that consider that every law is a command give to permission. Such theory is sometimes called the imperative theory of law.

In the explained sense nothing is obligatory if Rex has not commanded it, so that there are no situations that are obligatory for logical reasons. This means that in the Deontic Logic corresponding to this interpretation there cannot be any law of the form OA. I believe that this is the reason for Lemmon's elimination of R3 (if $\vdash A$ then $\vdash OA$) from his systems of Deontic and Epistemic Logic.

Systems of both types must always be situated at the left of our line *B*. This rejection implies also the rejection of A3 (Prior's principle) because it is of the proscribed form 'OA' and because, as has rightly been observed by Lemmon (1957), if it were added to the accepted principles, the rejected rule R3 could be derived.

If these comments are correct then the Deontic Logic for the *O*-operator and the *Pw*-operator (weakly permitted) is *C*2. If the *Ps*-operator (strongly permitted) is added, the axiomatic bases must be supplemented by

A2. 1 $\vdash OA \supset PsA$

and R5 for Ps, i.e.: If $\vdash A \supset B$ then $\vdash PsA \supset PsB$

III

These results may look somehow artificial, so I shall try to explain and delimit more sharply the domain of applicability of the interpretation we have dealt with.

'Op' and 'Psp' require for their truth a positive act of the person (Rex in our simplified society) or persons recognized as law-creating authorities. 'PwA', on the other hand, presupposes that the authorities have not done certain law-creating acts. By the way, this style of making the normative qualities dependent on the behavior of the authorities reflects the specific positive-law point of view. In this interpretation deontic sentences express descriptive propositions about prescriptive acts of the law-creating authorities.

When 'Op' is true Rex (or the corresponding authority as the case may be) has commanded that p, and this may be interpreted as meaning that what Rex has done is to say:

(1) 'It ought to be that p' (or 'p ought to be (done)')

When 'Psp' is true, what must occur is that Rex has said:

(II) 'It may be that p' (or 'p may be (done)')

Very frequently the sentences (I) and (II) have been taken as translations of the deontic sentences Op and Pp. I think that this interpretation is correct and very important. But we must notice that under it deontic sentences receive very different meanings from the ones previously discussed, because on it (I) and (II) do not have a descriptive but a prescriptive import. They are prescriptions; they are norms issued by the law-creating authorities.

I sum up these interpretations in Table I.

There are, at least, two alternatives for the logic of deontic expressions in the prescriptive interpretation. One may take D $[T(D)]$ or $D2$, but in both cases a restriction must be added in the formation-rules in order to exclude sentences in which a deontic operator occurs in the scope of an-

TABLE I

Deontic expression	Descriptive interpretation (Normative propositions)	Prescriptive interpretation (Norms)
Op	It is obligatory that p	It ought to be that p p ought to be (done)
$O \sim p$	It is forbidden (prohibited) that p It is obligatory that not p	It ought to be that not p p must not be (done)
Pp	It is permitted that p (with two interpretations: one for strong permission -Psp-, one for weak permission -Pwp-)	It may be that p p may be (done)

other deontic operator. The content of a norm, i.e., that affected by the norm-character expressed by the deontic operator, must be an action, an activity or a state of affairs that results as a consequence of an action or an activity, so that the expression that follows a deontic operator must be a *description* of one of these things, but it cannot be a prescription (a norm) itself. Deontic operators in the prescriptive interpretation generate norms (prescriptions) out of descriptions of a certain kind.

I take it that A1 and R4 are in this sense clearly logical principles for the following reasons. First, when someone prescribes that it ought to be that p and that it ought to be that q, the prescriptive import of what he has done is the same as if he were prescribing that it ought to be that p and q. Second, because the consequences of the prescribed situations are also prescribed; when someone prescribes that it ought to be that p he also indirectly prescribes that the consequences of p ought to be.

We saw in the descriptive interpretation that a state of affairs p is obligatory and forbidden when Rex (or the norm-creating authorities) has commanded that p and not p. I pointed out then that even if this is not an impossible situation it is a regrettable one. But why is it regrettable? Not because it is unfair, unjust or bad from some axiological point of view but because the authority has betrayed his prescriptive intention in prescribing too much (incompatible results). In this sense I believe that A2 represents a conceptual criterion for deontic consistency in the field of prescriptive (normative) discourse, so I understand that it must be accepted in the prescriptive normative interpretation.

We must also accept A0 as is shown by the following argument. Suppose that some subject asks some norm-authority whether he may produce p, then the authority may prescriptively disqualify the situation by saying either: 'You may not (do p)' or 'You ought to do not p' or 'You must not do p'.

'$\sim Pp$' and '$O \sim p$' have the same directive or prescriptive import, and so do 'Pp' and '$\sim O \sim p$'. A consequence of the acceptance of A0 and A2 is the acceptance of A2.1, which may be paraphrased by saying that in the prescriptive interpretation commanding implies authorizing.

If we wish to reconstruct prescriptive discourse as closely as possible to ordinary usage we must reject as ill-formed all those deontic sentences in which the expression that follows a deontic operator is contradictory or tautologous, i.e., all sentences of the form OA or PA in which A is a contradiction or a tautology, because in them there is no particular state of affairs to prescribe (command or authorize). In order to prescribe one must depict the commanded or authorized situation, and as Wittgenstein said, "Tautologies and contradictions are not pictures of reality. They do not represent any possible situations. For the former admit *all* possible situations, and the latter none" (Tr. 4.462). I think that the content of a norm must be a proposition in Wittgenstein's restricted sense in which "the proposition is only the *description* of a situation" (Notebooks 30 e). In this sense "Tautology and contradiction are not real propositions, but degenerate cases" (Ramsey p. 10). "Tautology and contradiction lack sense" (Tr. 4.461). They are the limiting points where proposition vanishes, "it is only by courtesy that they are called propositions" (Black, 229).

If we were to follow this line of thought the logical calculus would result unusually complicated. (Prof. Bulygin and I have developed things in this direction in 1971). So I have decided to be courteous and admit expressions of the form OA and PA where A is an expression of the indicated kind. Having taken this step I think that it is a matter of personal taste to adopt or not R3. Perhaps admitting it will give us a smoother calculus.

To sum up, the deontic logic for the prescriptive interpretation is $D2$ or $D[T(D)]$ with the restrictions in the rules of formation pointed out on pp. 453–454

We have now different logics that correspond to the descriptive and prescriptive interpretations of the deontic expressions. The essential points of difference lie around A2, A0 and A2.1. The three are valid only in the prescriptive interpretation. For the descriptive interpretation A2 is invalid, A2.1 is valid only in the strong meaning of P (Ps), and A0 is valid only in the weak meaning of P (Pw). In this my conclusions differ from Stenius' belief (1962) that $D\,[T(D)]$ is the logic that springs out of both the prescriptive and the descriptive interpretations.

IV

These two conceptions of deontic discourse are not unrelated. The descriptive point of view depends upon the prescriptive one in a sense that I shall try to explain following two different but parallel routes. The first consists in reconstructing the two interpretations in a mixed calculus developed in a single language, and the second will reconstruct the descriptive interpretation in the metalanguage of a prescriptively interpreted language.

Let us call O-language a language with the following characteristics: (1) an infinite list of variables, (2) the usual propositional connectives, (3) the O-deontic operator, (4) the set of wffs (O-formulas) is the union set of (i) the set of content-formulas (C-formulas), which is the smallest set that contains the variables and is closed with respect to the propositional connectives, (ii) the set of deontic formulas (D-formulas) which is the smallest set that contains all the formulas of the form OA when A is a C-formula, and is closed with respect to propositional connectives. In O-language we incorporate as axioms and rules of inference those of system $D\,[T(D)]$ replacing A0 by the definition $P = \sim O \sim$.

The O-language is understood in the prescriptive way. I shall refer as NO-language to an extension of the O-language obtained by adding an operator Nx, and whose set of wffs (NO-formulas) is the union set of the O-formulas and the set that contains all the formulas of the form NxA where A is a D-formula, and is closed under propositional operators. As the axioms and rules of inference[1] for the NO-language we take those of O-language plus:

AN: $\vdash (NxA \cdot NxB) \supset Nx(A \cdot B)$ Where A and B are D-formulas)
RN: If $\vdash A \supset B$ then $\vdash NxA \supset NxB$ (Where A and B are
                    D-formulas)

In *NO*-language we can introduce the following definitions:

D1 $\mathbb{O}p = NxOp$
D2 $\mathbb{P}sp = NxPp$
D3 $\mathbb{P}wp = \sim NxO \sim p$

These definitions are understood as the reconstruction of the deontic operators in the descriptive interpretation. With this interpretation we saw that 'It is obligatory that p' is true iff Rex has commanded that p, i.e., when *Rex has ruled* (issued a prescription, a norm, to the effect) *that p ought to be* (*done*). As 'p ought to be (done)' is formalized by Op (we must remember that O is used in the prescriptive way), 'Rex has ruled Op' finds its natural formalization in '$NxOp$' (where x stands for Rex, or the chosen authority as the case may be). This justifies $D1$ and the introduction of $\mathbb{O}p$ to reconstruct the descriptive interpretation of the obligatory-operator.

'It is permitted that p' in the strong sense is true iff Rex has authorized p, i.e., when *Rex has ruled that p may be* (*done*). As 'p may be (done)' is formalized by 'Pp', Rex has ruled that Pp is rendered by '$NxPp$' – $\mathbb{P}sp$ – which represents the descriptive interpretation of the strong permissive operator.

In order that 'It is permitted that p' be true in the weak sense it is necessary and sufficient that Rex has not commanded that not p, i.e., that is not true that Rex has ruled that it ought to be that not p. As '$O \sim p$' means that it ought to be that not p, so 'It is permitted that p' is represented by $\sim NxO \sim p (\mathbb{P}wp)$.

With this formalism (for the formal development of this approach see Alchourrón, 1969) the principles previously identified as the characteristics of the descriptive interpretation can be proved, that is:

Let A and B be two C-Formulas

$\vdash \mathbb{P}wA \equiv \sim \mathbb{O} \sim A$ (Corresponding to A0)
$\vdash (\mathbb{O}A \cdot \mathbb{O}B) \supset \mathbb{O}(A.B)$ (Corresponding to A1)
$\vdash \mathbb{P}w(A \vee B) \supset (\mathbb{P}wA \vee \mathbb{P}wB)$ (Corresponding to A1.1)
$\vdash \mathbb{O}A \supset \mathbb{P}sB$ (Corresponding to A2.1)
If $\vdash A \supset B$ then $\vdash \mathbb{O}A \supset \mathbb{O}B$ (Corresponding to R4)
If $\vdash A \supset B$ then $\vdash \mathbb{P}sA \supset \mathbb{P}sB$ (Corresponding to R5)

It can also be proved that the following formulas are not theorems:

(1) $\mathbb{P}sA \supset \sim \mathbb{O} \sim A$ (Corresponding to A0.1)

(2) $\sim \mathbb{O} \sim A \supset \mathbb{P}sA$ (Corresponding to the other half of A0)
(3) $\mathbb{O}A \supset \sim \mathbb{O} \sim A$ (Corresponding to A2)
(4) $\mathbb{O}A \supset \mathbb{P}wA$ (Corresponding to A2.1)
(5) $\mathbb{P}s(A \vee B) \supset (\mathbb{P}sA \vee \mathbb{P}sB)$ (Corresponding to A1.1)

From the fact that (1) is not a theorem it follows that '$\mathbb{P}sA . \mathbb{O} \sim A$' is consistent, i.e. that it is possible that it is permitted and forbidden that p. But when this happens we would say that the authority x has issued incompatible rules about p, so the following definition for 'x (the authority) has issued incompatible norms about p' may be introduced.

$$IN(A) \equiv (\mathbb{P}sA \cdot \mathbb{O} \sim A)$$

The possible falsehood of (3) shows the possibility for a state of affairs to be obligatory and prohibited, but it can be proved that when things are like that the authority x has issued incompatible norms. This is reflected in the theorem:

$$\vdash (\mathbb{O}A \cdot \mathbb{O} \sim A) \supset IN(A)$$

Formula (2) may be false but it may also be true. When it is true I will say that the authority x has determined a normative status for A ($DN(A)$), so I introduce the definition:

$$DN(A) \equiv (\mathbb{O} \sim A \vee \mathbb{P}sA)$$

The idea of normative determination may be used to characterize the concept of completeness for a system of norms or, as it is called in juristic parlance a system without gaps. The set of norms issued by the authority x is said to be complete when x has normatively determined every state of affairs.

So in *NO*-language we have two sets of deontic operators. On the one hand 'O' and 'P' which are interpreted in the normative or prescriptive way. On the other hand '$\mathbb{O}$', '$\mathbb{P}s$' and '$\mathbb{P}w$' that represent the descriptive interpretation, and whose logical principles depend upon the laws of the prescriptive operators and the norm-creation operator 'Nx' which is governed by the laws of system C2.

Some authors have thought that the logical principles of the deontic operators in both interpretations are the same. I take it that, when this has happened, what has occurred is that they were thinking of the ideal situation in which the source (or sources) of the norm has not created inconsistent rules and has determined a normative status for every action. Perhaps something like this is what happens with those who adopt the natural law or some ethical allcomprehensive point of view instead of the positive law point of view. This is even clearer in those writers who consider God to be the only source of morality and natural law principles.

Such a situation is formally reflected in the acceptance, as axiomatic principles, of:

$\vdash \sim IN(A)$ (Postulate of Universal Normative Consistency)
$\vdash DN(A)$ (Postulate of Completeness (of Universal Normative Determination))

With these two principles added to those previously presented, formulas (1)–(5) can be proved. Moreover the logic of the descriptive deontic operators turns out to be the same as the logic of the prescriptive deontic operators, and the difference between the two permissive descriptive operators vanishes, so that the two logics become isomorphic. I think that this is the reason why many authors talk of two interpretations of a single calculus instead of trying to build two formalisms for these two explicanda.

Up to this point we have introduced the descriptive operators and developed the logic of normative propositions, in the extended NO-language which belongs to the same language level as the O-language for norms.

V

A more illuminating approach is obtained if the logic of normative propositions is developed in the metalanguage of the language of norms. I shall now sketch the main features of this kind of analysis.

Let us construct an FO-language as an extension of the O-language in the following way:

(1) Besides the set of variables of the O-language we incorporate into the FO-language a different set of variables, F variables.

(2) The wffs are indentified through the following steps:

(i) let Clos (set of F variables) be the smallest set that contains the set of F variables and is closed with respect to the propositional connectives.

(ii) The set M of wffs is the union of: (a) the smallest set that contains Clos (set of F variables) and the set of D-formulas, and is closed with respect to the propositional connectives; and (b) the set of C-formulas.

(3) The set L of logical laws is the smallest subset of M that contains all propositional tautologies and logical laws of the O-language and is closed under the operation of detachment.

In the syntactical metalanguage of the FO-language we define, in Tarski's style, the set of consequences of a set α of sentences ($Cn(\alpha)$) as the smallest subset of M such that contains $\alpha + L$ and is closed under the operation of detachment.

We identify as the set of Cases (Ca) the set Clos (set of F variables) minus L; and as the set of (normative) contents (Con) the set of C-formulas minus L.

I shall use '-', '$\rightarrow$', 'o', '&' and '$\leftrightarrow$' for negation, material implication, inclusive disjunction, conjunction and material equivalence in the descriptive metalanguage of the normative FO-language, 'x' for an arbitrary element of Ca, and 'y' for an arbitrary element of Con. The connectives and operators of the object language are used in the metalanguage for autonymous reference.

In the metalanguage we can define:

D1a $\quad \mathbb{O}_\alpha(y/x) \leftrightarrow (x \supset Oy) \, \varepsilon \, Cn(\alpha);$

$\qquad\qquad\qquad\qquad$ Db1 $\quad \mathbb{O}_\alpha(y) \leftrightarrow Oy \, \varepsilon \, Cn(\alpha)$

D2a $\quad \mathbb{P}s_\alpha(y/x) \leftrightarrow (x \supset Py) \, \varepsilon \, Cn(\alpha);$

$\qquad\qquad\qquad\qquad$ D2b $\quad \mathbb{P}s_\alpha(y) \leftrightarrow Py \, \varepsilon \, Cn(\alpha)$

D3a $\quad \mathbb{P}w_\alpha(y/x) \leftrightarrow (x \supset O \sim y) \notin Cn(\alpha);$

$\qquad\qquad\qquad\qquad$ D3b $\quad \mathbb{P}w(y) \leftrightarrow O \sim y \notin Cn(\alpha)$

I explain now the intuitive background of this formal machinery. Let us choose α as the set of sentences that express all the norms issued (promulgated) by one or several law-creating authorities (for instance, Rex of our simplified construction). Normally Cn (α) minus L will be recognized as the normative system enacted by the chosen authority or authorities.

I believe also that lawyers would say that it is obligatory – in relation to α – to see to it that the state of affairs referred to by the sentence referred to by y occurs, i.e. $\mathbb{O}_\alpha(y)$, iff the norm Oy belongs to the system determined by α, i.e., when $Oy \, \varepsilon \, Cn \, (\alpha)$. And they would also say that it is obligatory – in relation to α – to see to it that the state of affairs referred to by the sentence referred to by y occurs in case the state of affairs referred to by the sentence referred to by x happens, i.e., $\mathbb{O}_\alpha(y/x)$ if and only if the norm Oy can be inferred from α and x, i.e. when $Oy \; \varepsilon \; \text{Cn}\,(\alpha+\{x\})$ or – what is equvalent – when $(x \supset Oy) \; \varepsilon \; \text{Cn}(\alpha)$. For example, it will be said that in Platina it is obligatory to pay in income tax of $ 10, in case of having an income of $ 1000 per annum if and only if the norm that says 'Those who have an income of $ 1000 per annum ought to pay an income tax of $ 10' is a consequence of the set of norms issued by Platinian authorities.

Mutatis mutandis it may be shown that definitions D2a, D2b, D3a and D3b reconstruct the descriptive meaning of strong and weak conditional and categorical permission respectively.

An important point must be noticed here. In the translations advanced for $\mathbb{O}_\alpha(y/x)$, $\mathbb{O}_\alpha(y)$, $\mathbb{P}s_\alpha(y/x)$, etc., we use metalanguage in a transposed way; we use it in what Carnap called the material mode of speech, in which "in order to say something about a word (or a sentence) we say instead something parallel about the object designated by the word (or the fact described by the sentence, respectively)" (Syntax, p. 309). Carnap detected very clearly this use of legal language when he wrote: "According to ordinary use of language, an action a of a certain person is called legal crime if the penal law of the country in which that person lives places the description of a kind of action to which a belongs in the list of crimes" (p. 308).

It may be proved that with the preceding definitions the logical laws for the descriptive interpretation of the deontic operators, and others for the conditional ones, can be obtained without any further addition. (For a more comprehensive development see Alchourron and Bulygin, 1971.)

I shall illustrate this point by presenting the characteristic principles of subalternation, distribution and correlation.

Conditional Operators

$\vdash \mathbb{O}_\alpha(y/x) \rightarrow \mathbb{P}s_\alpha(y/x)$

$\vdash \mathbb{O}_\alpha(y_1 \cdot y_2/x) \leftrightarrow \mathbb{O}_\alpha(y_1/x) \,\&\, \mathbb{O}_\alpha(y_2/x)$

$\vdash \mathbb{P}w_\alpha(y_1 \vee y_2/x) \leftrightarrow [\mathbb{P}w_\alpha(y_1/x) \circ \mathbb{P}w_\alpha(y_2/x)]$

$\vdash \mathbb{P}s_\alpha(y_1/x) \circ \mathbb{P}s_\alpha(y_2/x)] \rightarrow \mathbb{P}s_\alpha(y_1 \vee y_2/x)$

$\vdash \aleph(y/x_1 \vee x_2) \leftrightarrow [\aleph(y/x_1) \,\&\, \aleph(y/x_2)]$

[Where in the place of $\aleph$ is put on anyone of $\mathbb{O}_\alpha$, $\mathbb{P}s_\alpha$, or $\mathbb{P}w_\alpha$]

$\vdash \mathbb{O}_\alpha(y/x) \leftrightarrow -\mathbb{P}w_\alpha(\sim y/x)$
$\vdash \mathbb{P}w_\alpha(y/x) \leftrightarrow -\mathbb{O}_\alpha(\sim y/x)$
$\vdash \mathbb{O}_\alpha(y) \leftrightarrow \mathbb{O}_\alpha(y/x)$ for every x $[(x)\mathbb{O}_\alpha(y/x)]$
$\vdash \mathbb{P}s_\alpha(y) \leftrightarrow \mathbb{P}s_\alpha(y/x)$ for every x $[(x)\mathbb{P}s_\alpha(y/x)]$
$\vdash \mathbb{P}w_\alpha(y) \leftrightarrow \mathbb{P}w_\alpha(y/x)$ for some x $[(Ex)\mathbb{P}w_\alpha(y/x)]$

Categorical Operators

$\vdash \mathbb{O}_\alpha(y) \to \mathbb{P}s_\alpha(y)$
$\vdash \mathbb{O}_\alpha(y_1 \cdot y_2) \leftrightarrow [\mathbb{O}_\alpha(y_1) \& \mathbb{O}_\alpha(y_2)]$
$\vdash \mathbb{P}w_\alpha(y_1 \vee y_2) \leftrightarrow [\mathbb{P}w_\alpha(y_1) \circ \mathbb{P}w_\alpha(y_2)]$
$\vdash [\mathbb{P}s_\alpha(y_1) \circ \mathbb{P}s_\alpha(y_2)] \to \mathbb{P}s_\alpha(y_1 \vee y_2)$
$\vdash \mathbb{O}_\alpha(y) \leftrightarrow -\mathbb{P}w_\alpha(\sim y)$
$\vdash \mathbb{P}w_\alpha(y) \leftrightarrow -\mathbb{O}_\alpha(\sim y)$

In this approach the main metatheoretical issues concerning the consistency, completeness and independence of legal systems find their most natural presentation.

Universidad de Buenos Aires, Buenos Aires

NOTE

[1] The logic for 'Nx' is almost the same as the system A_0 of Rescher's assertion logic (Rescher, 1968, p. 277), the only difference being the absence in my system of the non-vacuousness postulate.

BIBLIOGRAPHY

Alchourrón, C. E., 1969, 'Logic of Norms and Logic of Normative Propositions', *Logique et Analyse* **47**, 242–68.
Alchourrón, C. E., and Bulygin, E., 1971, *Normative Systems*, Springer-Verlag, Wien-New York.
Black, M., 1964, *A Companion to Wittgenstein's Tractatus*, Cambridge University Press.
Carnap, R., 1937, *The Logical Syntax of Language*, K. Paul, Trench, Trubner & Co., London.
Hanson, W. H., 1965, 'Semantics for Deontic Logic', *Logique et Analyse* **31**, 177–91.
Hart, H. L. A., 1961, *The Concept of Law*, Clarendon Press. Oxford.
Lemmon, E. J., 1957, 'New Foundations for Lewis Modal Systems', *J. Symbolic Logic* **22**, 176–86.
Lemmon, E. J., 1965, 'Deontic Logic and the Logic of Imperatives', *Logique et analyse* **8**, 39–71.

Lemmon, E. J., 1966, 'Algebraic Semantics for Modal Logics', *J. Symbolic Logic* **31**, 191–218.
Ramsey, F. P., 1931, *The Foundations of Mathematics and Other Logical Essays*, Harcourt, Brace and Company, N.Y.
Rescher, N., 1968, *Topics in Philosophical Logic*, Reidel Publishing Company, Dordrecht, Holland.
Stenius, E., 1963, 'The Principles of a Logic of Normative Systems', *Acta Philosophica Fennica* **16**, 247–60.
von Wright, G. H., 1951b, *An Essay in Modal Logic*, North-Holland, Amsterdam.
von Wright, G. H., 1951a, 'Deontic Logic', *Mind* **60**, 1–15.
von Wright, G. H., 1967, 'Deontic Logics', *American Philosophical Quarterly* **4**, 136–43.
Wittgenstein, L., 1961, *Notebooks, 1914–1916*, Blackwell, Oxford.
Wittgenstein, L., 1922, *Tractatus Logico Philosophicus*, K. Paul, Trench, Trubner & Co., London.

PART IX

HISTORY OF PHILOSOPHY

HÉCTOR-NERI CASTAÑEDA

PLATO'S *PHAEDO* THEORY OF RELATIONS

Here is a fragment of an investigation into Plato's *Phaedo* General Theory of Forms, Relations, and Particulars. Here I report on two small related discoveries. First, I offer the historico-philosophical report that in the *Phaedo* Plato did put forward a theory of relations and relational facts, and I set forth that theory. Second, I offer the purely philosophical result that such a theory is logically sound and ontologically viable.

Contrary to the monolithic consensus among Plato scholars, in the *Phaedo* Plato did distinguish, and soundly, between relations and qualities, and dealt with genuine puzzles that arise in attempting to understand the nature of relational facts.[1] The reason why Plato's theory of relations has hitherto remained hidden to his commentators is this: his commentators have either not understood the nature of relations, or, more recently, they have adopted the dogma that a primary or simple relation is just one atomic or indivisible entity that generates facts by being instantiated at once by an ordered n-tuple. This is the view perspicuously represented by the standard notation '$A(x_1, ..., x_n)$' of the predicate calculus, where 'A' stands for an n-adic relation and each of 'x_1', ..., 'x_n' stands for an entity that has a fixed place in the ordered n-tuple $\langle x_1, ..., x_n \rangle$, which is the instance of A.[2] It might be suggested at this juncture that a nominalist as described by Quine does not countenance such entities as relations, because he does not quantify over predicates (or properties). Such a nominalist must, nevertheless, distinguish between a thing a being longer than another thing b, and the former being heavier than the second, and this distinction must lie in facts themselves, in nature. So, whether he quantifies over the grounds of the distinction or not, the truth of the case is that there are those grounds in nature, and those grounds are the atomic entities I am talking about.

Now, interestingly enough, it is astonishingly easy both to break away from the dogma of the uniqueness of the relational entity and to arrive at the proper appreciation of Plato's view. Thus, here we can enjoy the blending of a refurbishing historico-philosophical insight with a cathartic philosophical vision.

This essay is divided into five parts. Part I contains a fragment of my exegesis of the much maligned passage *Phaedo* 102B3–D3, in which Plato adumbrates a nice theory of relations. Part II formulates some principles of that theory. Part III discusses a trivial Platonic predicate calculus. Part IV contains the formulation of the appropriate semantics for a non-trivial Platonic predicate calculus. Since Plato's general theory of Forms is not presented here, I discuss only what I call Platonic relational structures. It is an easy matter to adapt for the Platonic predicate calculus a Henkin-type of completeness proof. Part V is the conclusion. The Appendix collects some of the most recent expressions of the well-entrenched view that Plato did not distinguish between relations and qualities at all, or not well enough.

I. TEXT AND EXEGESIS

Here I am not at all concerned with Plato's general theory of Forms. My only interest is to understand the theory of relational facts or propositions Plato is adumbrating in *Phaedo* 102B7–C4:

> Ἀλλὰ γάρ, ἦ δ' ὅς, ὁμολογεῖς τὸ τὸν Σιμμίαν
> ὑπερέχειν Σωκράτους οὐχ ὡς τοῖς ῥήμασι
> λέγεται οὕτω καὶ τὸ ἀληθὲς ἔχειν.
> οὐ γάρ που πεφυκέναι Σιμμίαν ὑπερέχειν
> τούτῳ τῷ Σιμμίαν εἶναι, ἀλλὰ τῷ μεγέθει ὃ
> τυγχάνει ἔχων. οὐδ' αὖ Σωκράτους ὑπερέχειν, ὅτι
> Σωκράτης ὁ Σωκράτης ἐστίν, ἀλλ' ὅτι σμικρότητα
> ἔχει ὁ Σωκράτης πρὸς τὸ ἐκείνου μέγεθος; Ἀλμθῆ.

The existing translations of this passage manage pretty well to convey the sense of the original Greek. What is needed is not so much another translation, but a philosophical exegesis. As a first step of the exegesis, we must unravel the ellipses of the text. I use square brackets to provide expressions that make the English clearer, angular brackets to supply ellipses in the Greek, and parentheses to furnish synonyms or preserve a Greek locution. Thus, a useful unliterary literal translation runs as follows:

B7 "But then," said he ⟨Socrates⟩, "do you accept [the following]?: That Simmias surpasses (is taller than) Socrates is not true as said with these words.

C1 For [it is] not in any way [the case that] Simmias has in himself (πεφυκέναι³ Σιμμίαν) to surpass by [virtue of] being Simmias, but [it is the case that ⟨Simmias has in himself to surpass⟩] by [virtue of] the tallness (τῷ μεγέθει) he happens to have (τυγχάνει ἔχων³);

C3 nor further [is it the case that ⟨Simmias has in himself⟩] to surpass Socrates because Socrates is Socrates, but ⟨Simmias has in himself to surpass Socrates⟩ because Socrates possesses shortness (σμικρότητα) with respect to his ⟨Simmias's⟩ tallness (μέγεθος)?"

In this passage Plato deals with the puzzle as to how the sentence 'Simmias is taller than Socrates' is not literally true, i.e., how it fails to perspicuously reveal the truth of the fact it expresses. He solves the puzzle in two steps. The first step makes two points that belong to the general theory of Forms and particulars: (i) surpassing in height involves the Form tallness; (ii) for an ordinary particular to surpass (another) it must have in itself a character tall. Here I will ignore (ii). The second step of Plato's solution also brings out two points: (iii) surpassing in height also involves the Form shortness; (iv) the fact that Simmias surpasses Socrates requires the joint or in-company participation of Simmias in tallness and of Socrates in shortness. Thus, the puzzle is solved. The sentence 'Simmias is taller than Socrates' does not reveal the truth it expresses perspicuously, because this sentence mentions only one Form, tallness, whereas the truth or fact in question involves two Forms, tallness and shortness. This is in a nutshell my exegesis of the passage.

To nail down my exegesis, I want to examine the opening puzzle. The statement that the words 'Simmias is taller than Socrates' do not reveal perspicuously the truth they express is itself very puzzling. Yet Plato has Socrates proceed directly to the explanation. Furthermore, he has Socrates ask as a matter of course for Simmias's acceptance of the puzzle, and he does not present Simmias as at all puzzled. This is also puzzling. Undoubtedly, Simmias knows something that went on before. In fact, a check of the preceding discussions explains everything. In 100C Plato formulates the crucial principle that ordinary things have the properties they have by participation in Forms. He then discusses several examples, and in

100E he has Socrates explain that "by virtue of tallness both tall things are tall and taller things are taller, and by virtue of shortness shorter things are shorter." This is it! If, like Simmias, we keep this point in mind, when we come to the sentence 'Simmias is taller than Socrates' we can see, just as immediately as Simmias saw it, that something is ontologically wrong with this sentence. The fact Simmias-being-taller-than-Socrates involves *both* Simmias's being taller and Socrates's being shorter with respect to each other; yet the sentence fails to mention the other component of the fact.

Plato makes Socrates repeat at 101A2–6 that taller things are taller by Tallness and shorter by Shortness. He iterates each point within the repetition. Furthermore, he rejects the explanation that a man is taller than another by virtue of a head as an ontological analysis of the fact in question. And his reason for rejecting it is most revealing:

> You would, I think, fear that someone may confront you with the retort, if you said that a man was taller than another, and this shorter, by a head, first that by *the same thing* the taller is taller and the shorter, and... [*Phaedo*, 101A5–B; my italics]

Thus, consider the fact *Simmias is taller than Socrates*. It is the same fact *Socrates is shorter than Simmias*. That is to say, that fact involves Simmias's being taller as well as Socrates being shorter – and, as the preceding passage shows, Plato holds that it is an error to be feared to claim that Simmias's tallerness (than Socrates) and Socrates's shorterness (than Simmias) consist of, or are analyzable as, the same thing. Since each involves a Form, there must, therefore, be two Forms involved in the fact *Simmias is taller than Socrates*.

Phaedo 100E5–6 applies to the relation taller-than the principle that all characters or aspects of ordinary particulars, hence, all facts involving ordinary particulars, are to be analyzed in terms of participation in Forms. *Phaedo* 101A2–8 repeats the point and adds the crucial principle that facts involving the relation taller-than involves two Forms, Tallness and Shortness. By implication Plato submits that taller-than is the pair {Tallness, Shortness}. *Phaedo* 102C1–D1 iterates the previous points and adds a third one: Tallness and Shortness are structured (Πρὸς) by a law of joint instantiation: a simple relational fact involving taller-than is a two-pronged fact. Thus, Plato's discussion exhibits an organic cumulative development of his conception of the structure of relational facts – at least dyadic relational facts.

To sum up, the fact unperspicuously expressed by the sentence 'Simmias is taller than Socrates' must be understood as involving: (1) the two Forms tallness and shortness, (2) participation in each Form by one person only, (3) a connection between the two Forms that requires that they be participated in simultaneously, and (4) a derivative connection between the two participating persons that reflects the connection between the two Forms.

II. PLATO'S PHAEDO THEORY OF RELATIONS

In the *Phaedo* Plato distinguishes three kinds of entities: (a) Forms, (b) forms or characters in ordinary particulars, and (c) ordinary particulars. Here I will ignore (b), because they belong in Plato's General Theory of Forms and Particulars. Thus, the outcome of the preceding exegesis is this:

P1 Ordinary particulars have the properties they have by participation in Forms.
P2 All Forms are monadic, i.e., each Form is instantiated only by one particular in each fact it is involved in: no Form is ever instantiated by pairs or other n-tuples, whether ordered or not.
P3 Some facts consist of a particular instantiating, or participating in, a Form: they are *single-pronged*. Other facts are *multiple-pronged*: they consist of an array of Forms each instantiated by one particular, where these instantiations do not by themselves constitute facts.
P4 Forms that can enter into multiple-pronged facts cannot enter into single-pronged facts. This is the *law of factual enchainment*. Forms governed by this law constitute Form-chains or *relations*.

For an example consider the Forms Tallness and Shortness. They constitute the chain Tallness-Shortness, which is the relation taller-than. Thus, the fact that Simmias is taller than Socrates can be perspicuously represented by writing "Tallness(Simmias)-Shortness(Socrates)."

In general, an n-adic relation is on the view adumbrated by Plato in the

Phaedo a set of *n* Forms governed by the law of factual enchainment. For further illustration consider the fact or proposition that Mary gave John the first copy of *Noûs*. According to the standard view, this is a case of the relation Giving being instantiated, or thought to be so, by the ordered triple ⟨Mary, John, the first copy of *Noûs*⟩. On the Phaedian theory, the relation Giving is a triple of Forms, say: *Givership, Giveeship*, and *Givenship*. Thus, the fact, or proposition, we are dealing with is

> Givership (Mary)-Giveeship (John)-Givenship (the first copy of *Noûs*).

Here we have a three-pronged fact or proposition. None of its three components *Givership (Mary), Giveeship (John)*, and *Givership (the first copy of Noûs)* is itself a proposition or fact.

In short, in the *Phaedo* Plato *does* adumbrate a reduction of relations to monadic Forms, but he does *not* propose a reduction of relational facts to monadic ones.

Now, is such a Platonic view wrong-headed as Plato scholars think it is? The answer is an emphatic "No." Indeed, there is a trivial interpretation of the predicate calculus that makes the "No" trivial and, consequently, absolutely correct. But there is a non-trivial, or not-so-trivial, interpretation that makes the "No" interesting, yet still emphatic.

III. TRIVIAL PLATONIC QUANTIFICATIONAL CALCULI

Plato's reduction of relations to monadic Forms is so logically correct that it can be trivially represented in the standard calculus of properties, i.e., the first-order quantificational calculus.

Take any formulation of the first-order functional calculus. It assumes a set of predicate symbols and a set of individual variables, and, perhaps, a set of individual constants. It has a rule of formation to the effect that a predicate symbol of degree *n* followed by *n* individual signs is a well-formed formula. To fix the idea, let $A^{n,j}$ be the *j*th predicate of degree *n*. Let the rule of formation in question be:

(r.1) Sequences of primitive signs of the calculus of the following form are wffs:
$A^{n,j}(x_1, ..., x_n)$,
where $x_1, ..., x_n$ are all individual signs.

PLATO'S 'PHAEDO' THEORY OF RELATIONS

Consider now a cumbersome typographical or spelling alteration of that calculus, as follows: each predicate symbol $A^{n,j}$ is written as the *one* symbol $A^{1,n,j}(\ )\text{-}A^{2,n,j}(\ )\text{-}...\text{-}A^{n,n,j}(\ )$. And the rule of formation replacing (r.1) is

(r.1a) Sequences of primitive signs of the calculus of the following from are wffs:
$A^{1,n,j}(x_1)\text{-}A^{2,n,j}(x_2)\text{-}...\text{-}A^{n,n,j}(x_n)$,
where $x_1, ..., x_n$ are all individual signs.

In the cumbersome calculus the *atomic* formulas (aside from identities, if any) are precisely of the cumbersome form illustrated in the text of (r.1a). Undoubtedly, these are typographically complex formulas; yet they are ontologically pleasing to the Phaedian Platonist. They reveal perspicuously the structure of relational facts.

Clearly, as long as we merely replace one symbol for another, we have not tampered with the semantics of the first-order functional calculus. Every single theorem and meta-theorem remains. Therefore, there is nothing in the alteration that can logically ashame the Phaedian Platonist.

IV. PLATONIC SEMANTICS: SUB-PLATONIC MODELS

The Platonist cannot be content with the syntactical change discussed above. He wants to confer some independence to each part of the one cumbersome symbol that represents a relation. That is to say, he cannot be happy with the mere typographical change, because he wants the complexity of the symbol to represent some ontological structure: he wants to assign a Form to each part $A^{i,n,j}$ ($i = 1, 2, \cdots, n$) of the relational symbol. This piece of semantics is indeed what makes his typographical change a genuine orthographic alteration, i.e., an ontologically correct writing.

The Platonist wants, therefore, to assume a reservoir of primitive predicate signs of the form $A^{i,n,j}$, which are to be interpreted as expressive of Forms. Thus, he wants a formation rule more profound than (r.1a), namely:

(R.1b) Sequences of primitive signs of the calculus of the following form are wffs:

$$A^{h_1,n,j}(x_1)A^{h_2,n,j}(x_2)\cdots A^{h_n,n,j}(x_n),$$
where: (a) $h_s \neq h_t$, if $s \neq t$; (b) $h_1 + h_2 + \cdots + h_n = 1 + 2 + \cdots n$; and (c) each of $x_1, x_2, \cdots$, and x_n is an individual sign.

This rule of formation allows each predicate $A^{i,n,j}$ an individuality or independence of its own. For one thing, the order of each predicate symbol in the atomic wff characterized by (R.lb) is not fixed. This leaves it open to allow the order of the Forms in a relation to make a difference to the relation. However, I am convinced that the theory of relations adumbrated in the *Phaedo* does not distinguish facts by the order in which Forms are instantiated. I feel confident that Plato considered the fact that Simmias is taller than Socrates to be identical with the fact that Socrates is shorter than Simmias. Part of my evidence is the symmetrical explanation he gives in the passage above exegesized. Thus, the Platonic theory includes an axiom of commutativity, namely

Ax. 1. $A^{h_1,n,j}(x_1)\cdots A^{h_i,n,j}(x_i)A^{h_{i+1},n,j}(x_{i+1})\cdots A^{h_n,n,j}(x_n) \supset$
$A^{h_1,n,j}(x_1)\cdots A^{h_{i+1},n,j}(x_{i+1})A^{h_i,n,j}(x_i)\cdots A^{h_n,n,j}(x_n)$.

The remaining axioms are identical with the cumbersome axioms of the trivially modified first-order functional calculus introduced in Section III.

Now, however, we cannot take for granted that the standard metatheorems of the first-order functional calculus hold. For one thing, any proof of consistency or completeness will have to reckon with the fact that each predicate $A^{i,n,j}$ is assigned a set of individuals in the domain of interpretation. Thus, we need a special kind of semantical model suitable to the Platonic first-order functional calculus.

As it will be recalled, for the standard quantificational calculus, a *standard model* is an ordered pair $\langle S, D \rangle$, where O is the null Set:

1. $S \neq O$.
2. $D \neq O$.
3. Every member of D is a set of ordered n-tuples $\langle r_1, \cdots r_n \rangle$ such that: (i) $n > 0$; (ii) each r_i is a member of S.

A *standard interpretation* I over a standard model is a function that assigns to each individual sign of the calculus a member of S and to each predicate $A^{n,j}$ of degree n a member of D whose members are n-tuples.

A *standard valuation* V determined by a (standard) interpretation I over a (standard) model m is a function that assigns *truth* or *falsity* to each closed wff. The fundamental atomic clause is this:

VI. $V(A^{n,j}(x_1, \cdots, x_n) I, m)$ is true, if and only if
$\langle I(x_1, m), \cdots, I(x_n, m) \rangle$ is a member of $I(A^{n,j,m})$.

We are not dealing here with Plato's whole Phaedian Theory of Forms, for which I have formulated suitable Platonic semantical models. These models take into account the third type of entity, namely, characters in things, that Plato discusses in the *Phaedo*. For the purely relational part of the theory we need only what I call Platonic relational structures. Their characteristic feature is that they present the structure of the extensions of relations.

Let 'O' represent the null set. Then a *Platonic relational structure* is an ordered pair $\langle S, F \rangle$, where:

1. $S \neq O$.
2. $F \neq O$.
3. Every member f of F is an ordered n-tuple $\langle r_1, \cdots, r_n \rangle$, such that: (i) $n > 0$; (ii) $r_i \neq O$, for $i = 1, \cdots, n$; (iii) each r_i is a map of some ordinal into S.

The interpretations that make the preceding structures suitable models for the Platonic first-order functional calculus above discussed are definable as follows:

Let $m = \langle S, F \rangle$ be a Platonic relational structure. Then an *interpretation* I over m of an individual sign x and a predicate symbol A of the Platonic calculus is a function stisfying these conditions:

11. $I(x, m)$ is a member of S.
12. $I(A, m)$ is a member of a member of F.

Each interpretation yields a valuation of the wffs of the calculus in the standard way, except for the case of atomic relational wffs, namely:

V1. $V(A^{h_1}(x_1) \cdots A^h n(x_n), I, m)$ is true, if and only if there is an ordinal kth such that the kth member of
$I(A^{h_i}, m) = I(x_i, m)$ for $i = h_1, \cdots, h_n$.

As is customary, a set β of wffs of the Platonic calculus is *satisfiable* if and only if there is a Platonic relational structure m and there are functions I and V such that for each member p of β, $V(p, I, m)$ is true.

It is a routine affair to construct both a consistency proof and a Henkin-type proof of the completeness of Platonic calculus with respect to Platonic relational structures. They follow easily from the metatheorem below.

First let us define the following syntactical function c that maps Platonic wffs onto standard wffs, where f and g are Platonic wffs:

1. $c[A^{h_1,n,j}(x_1)\cdots A^{h_n,n,j}(x_n)] = A^{n,j}(x_1,\cdots,x_n)$
2. $c[\sim f] = \sim c[f]$
3. $c[f\ k\ g] = c[f]\ k\ c\ [g]$, where k is any dyadic connective.
4. $c[Qf] = Qc[f]$, where Q is any quantifier.

Metatheorem. A Platonic wff f is satisfied by a Platonic structure, if and only if the corresponding standard formula $c[f]$ is satisfied by some standard model.

Proof. Obviously, the satisfaction of atomic wffs is crucial. Patently, given an atomic wff f and a model $m = \langle S, F, \rangle$ that satisfies f, we can construct a standard model $m' = \langle S, D \rangle$ that satisfies $c[f]$, and vice versa.

Let f be $A^{h_1,n,j}(x_1)\cdots A^{h_n,n,j}(x_n)$,

and let the model that satisfies f be $m = \langle S, D \rangle$ and let I be the interpretation involved. Then there is an ordinal d such that dth member of each $I(A^{i,n,j}, m)$ is $I(x_{i,m})$. Consider the standard interpretation I' and model m' such that $I'(x, m') = I(x, m)$ and $I'(A^{n,j}, m')$ is the set of all ordered n-tuples $\langle r_1,\cdots, r_n \rangle$ such that for some ordinal e the eth member of $I(A^{i,n,j}, m)$ is r_i. Let $m' = \langle S, D \rangle$, where D is the set of all such $I'(A^{n,j}, m')$ for all n and j. Clearly, $\langle I'(x_1, m'),\ldots, I'(x_n, m') \rangle$ is a member of $I'(A^{n,j}, m')$. Hence, $c[f]$ is satisfied by m'.

Now, if $c[f]$ is satisfied by a standard model $m' = \langle S, D \rangle$, by the well-ordering theorem (that is, the axiom of choice) each member of D can be well-ordered into a set D_n of ordered n-tuples $\langle r_1,\cdots, r_n \rangle$. Let $p^{i,n}$ be the ordinal map whose eth member is the element r^i of the eth n-tuple of D_n. Let p^n be the ordered n-tuple $\langle p^{1,n},\cdots, p^{n,n} \rangle$ for each n. Let P be the set of all such p^n's. Then the structure $m = \langle S, P \rangle$ is a Platonic model that satisfies f.

V. CONCLUSION

I am pleased to have been able to vindicate Plato from the oft-rehearsed charge of not having distinguished relations from qualities. Not only does

Phaedo 102B7–C4 show quite clearly that he did make the proper distinction, but the theory of relations he adumbrated there is logically sound and ontologically viable. Furthermore, it is refreshing to think of relations not as Forms or universals, but as chains of ontologically tied universals.

Naturally, now that we have a clear understanding of Plato's *Phaedo* theory of relations and relational facts there is plenty of work to do. We must examine the other dialogues for alterations or even preservation of that theory. Moreover, there are those arguments of Aristotle that purport to reduce Plato's Theory of Forms to absurdity on account of relations. But of this I shall say more at some other time.

APPENDIX

The view that Plato had a confused conception of relations and failed to distinguish them from qualities has been endorsed by the following most distinguished among Plato scholars:

(a) R. C. Cross and (b) A. D. Woozley: "This seems to be the gist of the argument here, and if so, this *Theaetatus* passage *again* [my emphasis] illustrates Plato's failure to distinguish between two differing sorts of concepts, relational and non-relational, and the unreal puzzles that are consequently generated," on *Plato's Republic: A Philosophical Commentary* Macmillan, London, Melbourne, Toronto; St. Martin's Press, New York, 1966), p. 157; (c) Julius E. Moravcsik: "This is not the same distinction as the one drawn by modern philosophers between [non-relational] properties and relations. This latter distinction is not made either in the *Phaedo* or in the *Sophist*," on 'Being and Meaning in the *Sophist*', *Acta Philosophica Fennica* **14** (1962), p. 54nl.; (d) Francis M. Cornford: "In the whole argument [in *Phaedo* 100C1–102D3] no distinction is drawn between qualities and relations. Tallness is treated as if it were a quality like whiteness, inherent in the tall person, but with the peculiarity that he has it 'towards' or 'in comparison with' (πρός) the shortness of another person" from his *Plato and Parmenides* (Bobbs-Merrill, Indianapolis, Indiana: 1956), p. 78. As we shall see that peculiarity makes the whole difference of the world between consistency and inconsistency; (e) G. E. L. Owen who undoubtedly means to include Plato in his sweeping indictment in his 'A Proof in the ΠΕΡΙ ΙΔΕΩΝ', *Journal of Hellenic Studies* **77** (1957), p. 110: "One aim of the second part of the *Parmenides*,

I take it, is to find absurdities in a similar treatment of 'one'. It is the extreme case of Greek *mistreatment* [my italics] of 'relative terms in the attempt to assimilate them to simple adjectives[40]," where the footnote refers to footnote 1 pertaining to the passage from Cornford quoted in (d); (f) R. Hackforth, who does attribute to Plato a momentary semi-awareness of the distinction: "Plato has labored the point about opposites excluding opposites... he thought it was *prima facie* incompatible with the phenomena of the short man who is also tall... this incompatibility... is illusory, since 'tall' and 'short' are not qualities, but relations; or to put it in other words, [H] Simmias does not in fact contain two forms which he presents to Socrates and Phaedo respectively, but only one (relevant) form, i.e., character, namely *stature*, which remains unchanged whatever bystanders there may be. The curious thing is that Plato appears to be at least on the verge of realizing this... Yet this *semi-awareness* [my italics] of the distinction between qualities and relations is, it seems, only momentary; from 102D5 onwards it disappears," on Plato's *Phaedo* (Liberal Arts, The Bobbs-Merrill Company, Indianapolis-New York 1955), p. 155.

The statement I have labeled 'H' is, apparently, Hackforth's own statement of the matter. But as I argue in Section I of the preceding essay, it is not Plato's view and it is *not* what, in spite of Hackforth's next claim, Plato was "on the verge of realizing." In short, Hackforth really lends Plato no service when he credits Plato with a "momentary semi-awareness" of H. As I show in Sections III and IV of the preceding essay, Plato has such a nice conception of relations that he stands in no need of such charitably-meant commentaries as Hackforth's.

The other recent translator and commentator of the *Phaedo*, R. S. Bluck, simply says nothing about Plato's conception of relations. In his *Plato's Phaedo* (Library of Liberal Arts, The Bobbs-Merrill Company, Indianapolis-New York, 1955) p. 118, he merely summarizes Plato's statement about comparisons: "Opposite qualities, then, cannot coalesce with one another, although they can sometimes exist *side by side*, as it were, in sensible objects, when they are predicated of them in different respects or different relationships (if, for example, A is short by comparison with B, and tall by comparison with C, 102C)."

Like anybody else, Plato has, of course, a use for sentences of the form 'A is short (tall) with respect to C', or 'A is shorter (taller) than B'. But his problem in *Phaedo* 102B3–D3 is neither to use them nor to remind the reader of their use. His is the serious problem of providing in terms of his theory of Forms, an ontological analysis of what is expressed by sentences of that form. Bluck does not even mention that this is Plato's task there.

Indeed, the view that not only Plato but most philosophers up to the nineteenth century have grossly misunderstood the nature of relations is actually very common. Thus, Cornford found it easy to indict Plato and Aristotle by claiming that "it was reserved for still living logicians [apparently, A. N. Whitehead and B. Russell] to discover that a proposition like 'Socrates is shorter than Phaedo' has two subjects with a relation between them, and no predicate at all," *Plato's Theory of Knowledge* (The Liberal Arts Press, Inc., New York, 1957), pp. 28–36. Likewise, Bertrand Russell implies that relations were never really understood until after Hegel, when on page 158 of his *A History of Western Philosophy* (Simon and Schuster, New York, 1945), he asserts: "The idea of a relational proposition seems to have puzzled Plato, as it did most of the great philosophers down to Hegel (inclusive)."

Very much the same view is adopted as a matter of course by a most recent writer. On page 35f of his *Bradley's Metaphysics and the Self* (Yale University Press, New Haven and London, 1970), Garrett L. Vander Veer writes: "Bradley's puzzles over relations are not peculiar to him. In fact, classical substance-oriented philosophers were extremely perplexed about relations. For example, in the *Phaedo*, Plato tried to assimilate relations to the model that he has already given for qualities, whereas Leibniz' metaphysics reflects his attempt to reduce relations to qualities. Neither attempt was noticeably successful." This judgement on the *Phaedo* is at least insightful in attributing Plato a deliberate awareness of the difference between qualities and relations. It is also historically correct in that apparently the success of Plato's reduction of relations to qualities has not been noticed.

My contention in the preceding essay is, naturally, that Plato's puzzlement about relations culminated, in the *Phaedo*, precisely, in the adumbration of a logically viable theory of relations, which successfully reduces relations to monadic Forms. The metaphysical feasibility of the theory is at least on exactly the same footing and level as the metaphysical feasibility of any theory that postulates the existence of universals of quality. And it is even *more* feasible, if Russell's claim is correct that "As a matter of fact, if any one were anxious to deny altogether that there are such things as universals, we should find that we cannot strictly prove that there are such entities as *qualities*,... whereas we can prove that there must be *relations*," *The Problems of Philosophy* (Oxford University Press, London, reset edition of 1946), p. 95.

Indiana University

NOTES

[1] For endorsements of the well-entrenched view that Plato failed to make the proper distinction between qualities and relations see the Appendix to this essay.

[2] An extreme subservience to a standard notation for dyadic relations led Cornford to a most unjustified castigation of Plato for his discussion in *Phaedo* 102B3-D3: "Obviously, the author of the *Categories* did not conceive of relations as subsisting between two things, as they are now symbolized by R standing between a and b in aRb. ... It was reserved for still living logicians to discover that a proposition like 'Socrates is shorter than Phaedo' has two subjects with a relation between them, and no predicate at all. ... That Plato conceived relative terms in the same way is clear from the *Phaedo,* where he speaks of a man partaking of tallness in the same way that he partakes of beauty." This appears on pages 283f of his Plato's *Theory of Knowledge*. (The Liberal Arts Press, Inc., New York, 1957). It should be obvious to the reader of my essay below how wrong I think this commentary by Cornford is, not only in this interpretation of the *Phaedo*, but also on the nature of relations.

[3] It seems to me obvious that those commentators who find in the pair of locutions πεφυκέναι and τυγχάνει. an expression of the contrast between essential and accidental properties are doing serious and irrelevant violence to the text. Among such commentators are Bluck, *Plato's Phaedo* Library of Liberal Arts published by the Bobbs-Merrill Company, (Indianapolis-New York, 1955), and Robin, *Platon: Oeuvres Complètes*, Vol. I (Bibliotèque de la Pléiade, Paris, 1966). In fairness to Hackforth, it must be trumpeted that he explicitly rejects that interpretation of the pair of locutions.

SYNTHESE LIBRARY

Monographs on Epistemology, Logic, Methodology,
Philosophy of Science, Sociology of Science and of Knowledge, and on the
Mathematical Methods of Social and Behavioral Sciences

Editors:

DONALD DAVIDSON (The Rockefeller University and Princeton University)
JAAKKO HINTIKKA (Academy of Finland and Stanford University)
GABRIËL NUCHELMANS (University of Leyden)
WESLEY C. SALMON (Indiana University)

ROBERT S. COHEN and MARX W. WARTOFSKY (eds.), *Boston Studies in the Philosophy of Science.* Volume IX: *A. A. Zinov'ev: Foundations of the Logical Theory of Scientific Knowledge (Complex Logic).* Revised and Enlarged English Edition with an Appendix by G. A. Smirnov, E. A. Sidorenka, A. M. Fedina, and L. A. Bobrova. 1973, XXII + 301 pp. (Cloth) Dfl. 72,—
(Paper) Dfl. 43,—

K. J. J. HINTIKKA, J. M. E. MORAVCSIK, and P. SUPPES (eds.), *Approaches to Natural Language. Proceedings of the 1970 Stanford Workshop on Grammar and Semantics.* 1973, VIII + 526 pp. (Cloth) Dfl. 115,—
(Paper) Dfl. 55,—

WILLARD C. HUMPHREYS, JR. (ed.), *Norwood Russell Hanson: Constellations and Conjectures.* 1973, X + 282 pp. Dfl. 65,—

MARIO BUNGE, *Method, Model and Matter.* 1973, VII + 196 pp. Dfl. 45,—

MARIO BUNGE, *Philosophy of Physics.* 1973, IX + 248 pp. Dfl. 57,—

LADISLAV TONDL, *Boston Studies in the Philosophy of Science.* Volume X: *Scientific Procedures.* 1973, XIII + 268 pp. (Cloth) Dfl. 61,50
(Paper) Dfl. 35,—

SÖREN STENLUND, *Combinators, λ-Terms and Proof Theory.* 1972, 184 pp. Dfl. 40,—

DONALD DAVIDSON and GILBERT HARMAN (eds.), *Semantics of Natural Language.* 1972, X + 769 pp. (Cloth) Dfl. 95,—
(Paper) Dfl. 45,—

MARTIN STRAUSS, *Modern Physics and Its Philosophy. Selected Papers in the Logic, History, and Philosophy of Science.* 1972, X + 297 pp. Dfl. 80,—

‡STEPHEN TOULMIN and HARRY WOOLF (eds.), *Norwood Russell Hanson: What I Do Not Believe, and Other Essays,* 1971, XII + 390 pp. Dfl. 90,—

‡ROBERT S. COHEN and MARX W. WARTOFSKY (eds.), *Boston Studies in the Philosophy of Science.* Volume VIII: *PSA 1970. In Memory of Rudolf Carnap* (ed. by Roger C. Buck and Robert S. Cohen). 1971, LXVI + 615 pp. (Cloth) Dfl. 120,—
(Paper) Dfl. 60,—

‡Yehosua Bar-Hillel (ed.), *Pragmatics of Natural Languages*. 1971, VII + 231 pp. Dfl. 50,—

‡Robert S. Cohen and Marx W. Wartofsky (eds.), *Boston Studies in the Philosophy of Science*. Volume VII: *Milič Čapek: Bergson and Modern Physics*. 1971, XV + 414 pp. Dfl. 70,—

‡Carl R. Kordig, *The Justification of Scientific Change*. 1971, XIV + 119 pp. Dfl. 33,—

‡Joseph D. Sneed, *The Logical Structure of Mathematical Physics*. 1971, XV + 311 pp. Dfl. 70,—

‡Jean-Louis Krivine, *Introduction to Axiomatic Set Theory*. 1971, VII + 98 pp. Dfl. 28,—

‡Risto Hilpinen (ed.), *Deontic Logic: Introductory and Systematic Readings*. 1971, VII + 182 pp. Dfl. 45,—

‡Evert W. Beth, *Aspects of Modern Logic*. 1970, XI + 176 pp. Dfl. 42,—

‡Paul Weingartner and Gerhard Zecha, (eds.), *Induction, Physics, and Ethics, Proceedings and Discussions of the 1968 Salzburg Colloquium in the Philosophy of Science*. 1970, X + 382 pp. Dfl. 65,—

‡Rolf A. Eberle, *Nominalistic Systems*. 1970, IX + 217 pp. Dfl. 42,—

‡Jaakko Hintikka and Patrick Suppes, *Information and Inference*. 1970, X + 336 pp. Dfl. 60,—

‡Karel Lambert, *Philosophical Problems in Logic. Some Recent Developments*. 1970, VII + 176 pp. Dfl. 38,—

‡P. V. Tavanec (ed.), *Problems of the Logic of Scientific Knowledge*. 1969, XII + 429 pp. Dfl. 95,—

‡Robert S. Cohen and Raymond J. Seeger (eds.), *Boston Studies in the Philosophy of Science*. Volume VI: *Ernst Mach: Physicist and Philosopher*. 1970, VIII + 295 pp. Dfl. 38,—

‡Marshall Swain (ed.), *Induction, Acceptance, and Rational Belief*. 1970, VII + 232 pp. Dfl. 40,—

‡Nicholas Rescher et al. (eds.), *Essays in Honor of Carl G. Hempel. A Tribute on the Occasion of his Sixty-Fifth Birthday*. 1969, VII + 272 pp. Dfl. 50,—

‡Patrick Suppes, *Studies in the Methodology and Foundations of Science. Selected Papers from 1911 to 1969*. 1969, XII + 473 pp. Dfl. 72,—

‡Jaakko Hintikka, *Models for Modalities. Selected Essays*. 1969, IX + 220 pp. Dfl. 34,—

‡D. Davidson and J. Hintikka (eds.), *Words and Objections: Essays on the Work of W. V. Quine*. 1969, VIII + 366 pp. Dfl. 48,—

‡J. W. Davis, D. J. Hockney and W. K. Wilson (eds.), *Philosophical Logic*. 1969, VIII + 277 pp. Dfl. 45,—

‡Robert S. Cohen and Marx W. Wartofsky (eds.), *Boston Studies in the Philosophy of Science*, Volume V: *Proceedings of the Boston Colloquium for the Philosophy of Science 1966/1968*, VIII + 482 pp. Dfl. 60,—

‡ROBERT S. COHEN and MARX W. WARTOFSKY (eds.), *Boston Studies in the Philosophy of Science*. Volume IV: *Proceedings of the Boston Colloquium for the Philosophy of Science 1966/1968*. 1969, VIII + 537 pp. Dfl. 72,—

‡NICHOLAS RESCHER, *Topics in Philosophical Logic*. 1968, XIV + 347 pp. Dfl. 70,—

‡GÜNTHER PATZIG, *Aristotle's Theory of the Syllogism. A Logical-Philological Study of Book A of the Prior Analytics*. 1968, XVII + 215 pp. Dfl. 48,—

‡C. D. BROAD, *Induction, Probability, and Causation. Selected Papers*. 1968, XI + 296 pp. Dfl. 54,—

‡ROBERT S. COHEN and MARX W. WARTOFSKY (eds.), *Boston Studies in the Philosophy of Science*. Volume III: *Proceedings of the Boston Colloquium for the Philosophy of Science 1964/1966*. 1967, XLIX + 489 pp. Dfl. 70,—

‡GUIDO KÜNG, *Ontology and the Logistic Analysis of Language. An Enquiry into the Contemporary Views on Universals*. 1967, XI + 210 pp. Dfl. 41,—

*EVERT W. BETH and JEAN PIAGET, *Mathematical Epistemology and Psychology*. 1966, XXII + 326 pp. Dfl. 63,—

*EVERT W. BETH, *Mathematical Thought. An Introduction to the Philosophy of Mathematics*. 1965, XII + 208 pp. Dfl. 37,—

‡PAUL LORENZEN, *Formal Logic*. 1965, VIII + 123 pp. Dfl. 26,—

‡GEORGES GURVITCH, *The Spectrum of Social Time*. 1964, XXVI + 152 pp. Dfl. 25,—

‡A. A. ZINOV'EV, *Philosophical Problems of Many-Valued Logic*. 1963, XIV + 155 pp. Dfl. 32,—

‡MARX W. WARTOFSKY (ed.), *Boston Studies in the Philosophy of Science*. Volume I: *Proceedings of the Boston Colloquium for the Philosophy of Science, 1961–1962*. 1963, VIII + 212 pp. Dfl. 26,50

‡B. H. KAZEMIER and D. VUYSJE (eds.), *Logic and Language. Studies dedicated to Professor Rudolf Carnap on the Occasion of his Seventieth Birthday*. 1962, VI + 256 pp. Dfl. 35,—

*EVERT W. BETH, *Formal Methods. An Introduction to Symbolic Logic and to the Study of Effective Operations in Arithmetic and Logic*. 1962, XIV + 170 pp. Dfl. 35,—

*HANS FREUDENTHAL (ed.), *The Concept and the Role of the Model in Mathematics and Natural and Social Sciences. Proceedings of a Colloquium held at Utrecht, The Netherlands, January 1960*. 1961, VI + 194 pp. Dfl. 34,—

‡P. L. GUIRAUD, *Problèmes et méthodes de la statistique linguistique*. 1960, VI + 146 pp. Dfl. 28,—

*J. M. BOCHEŃSKI, *A Precis of Mathematical Logic*. 1959, X + 100 pp. Dfl. 23,—

SYNTHESE HISTORICAL LIBRARY

Texts and Studies
in the History of Logic and Philosophy

Editors:

N. KRETZMANN (Cornell University)
G. NUCHELMANS (University of Leyden)
L. M. DE RIJK (University of Leyden)

LEWIS WHITE BECK (ed.), *Proceedings of the Third International Kant Congress*. 1972, XI + 718 pp. Dfl. 160,—

‡KARL WOLF and PAUL WEINGARTNER (eds.), *Ernst Mally: Logische Schriften*. 1971, X + 340 pp. Dfl. 80,—

‡LEROY E. LOEMKER (ed.), *Gottfried Wilhelm Leibnitz: Philosophical Papers and Letters*. A Selection Translated and Edited, with an Introduction. 1969, XII + 736 pp.
Dfl. 125,—

‡M. T. BEONIO-BROCCHIERI FUMAGALLI, *The Logic of Abelard*. Translated from the Italian. 1969, IX + 101 pp. Dfl. 27,—

Sole Distributors in the U.S.A. and Canada:
*GORDON & BREACH, INC., 440 Park Avenue South, New York, N.Y. 10016
‡HUMANITIES PRESS, INC., 303 Park Avenue South, New York, N.Y. 10010

LIBRARY OF DAVIDSON CO